T0276560

New Frontiers in Fuzzy Controllers

New Frontiers in Fuzzy Controllers

Edited by **Ron Nucci**

LANRYE
INTERNATIONAL

New Jersey

Published by Clanrye International,
55 Van Reypen Street,
Jersey City, NJ 07306, USA
www.clanryeinternational.com

New Frontiers in Fuzzy Controllers
Edited by Ron Nucci

© 2015 Clanrye International

International Standard Book Number: 978-1-63240-378-0 (Hardback)

This book contains information obtained from authentic and highly regarded sources. Copyright for all individual chapters remain with the respective authors as indicated. A wide variety of references are listed. Permission and sources are indicated; for detailed attributions, please refer to the permissions page. Reasonable efforts have been made to publish reliable data and information, but the authors, editors and publisher cannot assume any responsibility for the validity of all materials or the consequences of their use.

The publisher's policy is to use permanent paper from mills that operate a sustainable forestry policy. Furthermore, the publisher ensures that the text paper and cover boards used have met acceptable environmental accreditation standards.

Trademark Notice: Registered trademark of products or corporate names are used only for explanation and identification without intent to infringe.

Printed in the United States of America.

Contents

Preface

The world is advancing at a fast pace like never before. Therefore, the need is to keep up with the latest developments. This book was an idea that came to fruition when the specialists in the area realized the need to coordinate together and document essential themes in the subject. That's when I was requested to be the editor. Editing this book has been an honour as it brings together diverse authors researching on different streams of the field. The book collates essential materials contributed by veterans in the area which can be utilized by students and researchers alike.

Advances in the sphere of fuzzy controllers have been rapid, giving rise to the need for literature that documents all the recent advances both in theory and in applications. This book has been published with an aim of serving the same purpose. It is a compilation of various research outcomes on diverse applications of fuzzy control systems. At the center of numerous engineering challenges is the question of controlling different systems. The wide spectrum of these structures includes a variety of systems varying from the inverted pendulum to auto-focusing system of a digital camera. Fuzzy control systems have displayed their improved output in all these areas. This book examines new areas of fuzzy applications such as a two-layered control system, cycle-to-cycle fuzzy logic control etc. It will be a good source of reference for researchers, engineers, and postgraduate students specializing in associated fields of fuzzy control systems.

Each chapter is a sole-standing publication that reflects each author's interpretation. Thus, the book displays a multi-facetted picture of our current understanding of application, resources and aspects of the field. I would like to thank the contributors of this book and my family for their endless support.

Editor

Enhancing Fuzzy Controllers Using Generalized Orthogonality Principle

Nora Boumella, Juan Carlos Figueroa and Sohail Iqbal

Additional information is available at the end of the chapter

1. Introduction

In the early days, the parameters of the fuzzy logic systems were fixed arbitrary, thus leading to a large number of possibilities for FLSs. In 1992, it has been shown that linguistic rules can be converted into Fuzzy Basis Functions (FBFs), and numerical rules and its associated FBFs must be extracted from numerical data training. Since that time, a multitude of design methods to construct a FLS are proposed. Some of these methods are intensive on data analysis, some are aimed at computational simplicity, some are recursive and others are offline, but all based on the the same idea: *tune the parameters of a FLS using the numerical training data*. Methods for designing FLSs can be classified into two major categories: A first category where shapes and parameters of the antecedent MFs are fixed ahead of time and training data are used for tuning the consequent parameters, and a second category that consists of fixing the shapes of the antecedent and consequent MFs using training data to tune the antecedent and the parameters of the consequent.

Two kinds of FLSs, the Mamdani and the Takagi-Sugeno-Kang (TSK) FLSs are widely used and they are currently adopted by the scientific community. They solely differ in the way the consequent structure is defined. The fact that a TSK FLS does not require a time-consuming defuzzification process makes it far more attractive for most of applications.

In this chapter, we consider the first category to design a TSK FLS basing on alinear method. Our design approach requires a set of input-output numerical data training pairs. Given linguistic rules of the FLS, we expand this FLS as a series of FBFs that are functions of the FLS inputs. We use the input training data to compute these FBFs. Therefore, the system becomes linear in the FLS consequent parameters, and we consider each set of FBFs as a basis vector which is easy to be optimized. Then follows the consequent parameters optimization via a minimizing process of the error vector - *the output training data minus the FBFs vectors weighted by the consequent parameters* - norm. This minimzation can be obtained by applying the *Generalized Orthogonality Principle* (GOP). Optimization process is carefully analyzed in this chapter and its applications in two major areas of concern are demonstrated including robotics and dynamic systems. Firstly, we shall show the improved results with analysis upon the application of GOP in the Fuzzy Logic Controller (FLC) for an inverted pendulum. Secondly, we show how a FLS based on this principle enhances the performance of forecaster for the chaotic time series.

2. Fuzzy Logic Systems (FLS) basic concepts

2.1. Fuzzy sets

A *Fuzzy Set* (FS), $F \in X$ is a set of ordered pairs of a generic element x and its degree, namely *Membership Function* (MF), $\mu_F(x)$. Any FS can be represented as follows:

$$F = \{(x, \mu_F(x)) \,|\, \forall x \in X\} \tag{1}$$

where the membership degree of x, $\mu_F(x)$, is constrained to be betwwen 0 and 1 for all $x \in X$.

2.2. Mamdani FLS

An FLS is an intuitive and numerical system that maps crisp (deterministic) inputs to a crisp output. It is composed of four elements which are depicted in Figure 1. To completly describe this FLS, we need a mathematical formula that maps the crisp input $\mathbf{x}$ into a crisp output $y = f(\mathbf{x})$, we can obtain this formula by following the signal $\mathbf{x}$ through the fuzzifier to the inference block and into the defuzzifier. We explain, in this section, the working principle of this formula.

2.2.1. Rules

The FLS is associated with a set of *IF-THEN* rules with meaningful linguistic interpretations. The lth rule of a FLS having p inputs $x_1, ..., x_p$ and one output $y \in Y$, *Multiple Input Single Output* (MISO), is expressed as:

$$R^l : \text{If } x_1 \text{ is } F_1^l \text{ and, ... , and } x_p \text{ is } F_p^l \text{ THEN } y \text{ is } G^l \tag{2}$$

where F_i^l $(i = 1, 2, ..., p)$ are fuzzy antecedent sets wich are represented by their MFs $\mu_{F_i^l}$, and G^l is a consequent set where $l = 1, ..., M$ (M is the number of rules in the FLS).

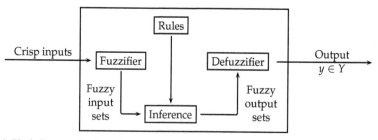

Figure 1. Block diagram of a fuzzy logic system

2.2.2. Fuzzifier

A fuzzifier maps any crisp input $\mathbf{x} = (x_1, ..., x_p)^T \in X_1 \times \cdots \times X_p \equiv \mathbf{X}$ into a fuzzy set $F_\mathbf{x}$ in $\mathbf{X}$ [8].

2.2.3. Inference

A fuzzy inference engine combines rules from the fuzzy rule base and gives a mapping from input fuzzy sets in $\mathbf{X}$ to output sets in Y. Each rule is interpreted as a fuzzy implication, i.e., a fuzzy set in $\mathbf{X} \times Y$, and can be expressed as:

$$R^l : F_1^l \times ... \times F_p^l \longrightarrow G^l = A^l \longrightarrow G^l \quad l = 1, ..., M \tag{3}$$

Usually in Mamdani FLS, the implication is replaced by a *t-norm*, i.e. (product or min). Multiple antecedents are connected by a t-norm, so a rule can be expressed by its MF as follows:

$$\mu_{R^l}(\mathbf{x}, y) = \mu_{F_1^l \times F_2^l \times ... \times F_p^l}(x_1, x_2, .., x_p) \bigstar \mu_{G^l}(y)$$
$$= \left[T_{i=1}^p \mu_{F_i^l}(x_i) \right] \bigstar \mu_{G^l}(y) \tag{4}$$

where T and $\bigstar$ are $t - norm$ operators (*product* or *min*). The p-dimensional input to R^l is given by the fuzzy set $A_\mathbf{x}$ whose MF is expressed as [8]

$$\mu_{A_\mathbf{x}}(\mathbf{x}) = \mu_{X_1}(x_1) \bigstar ... \bigstar \mu_{X_p}(x_p) = T_{i=1}^p \mu_{X_i}(x_i) \tag{5}$$

Each rule detemines a fuzzy set B^l in Y which is derived from the sup $- \bigstar$ composition. Then, the MF of this output set is expressed as [8]

$$\mu_{B^l}(y) = \mu_{A_\mathbf{x} \circ R^l}(y) = \sup_{\mathbf{x} \in \mathbf{X}} \left[\mu_{A_\mathbf{x}}(\mathbf{x}) \bigstar \mu_{R^l}(\mathbf{x}, y) \right] \tag{6}$$

$$\mu_{B^l}(y) = \sup_{\mathbf{x} \in \mathbf{X}} \left[T_{i=1}^p \mu_{X_i}(x_i) \bigstar \left(\left[T_{i=1}^p \mu_{F_i^l}(x_i) \right] \bigstar \mu_{G^l}(y) \right) \right] \tag{7}$$

Finally, the lth rule is expressed as follows

$$\mu_{B^l}(y) = \mu_{G^l}(y) \bigstar \left[T_{i=1}^p \mu_{F_i^l}(x_i) \right] \quad y \in Y \tag{8}$$

2.2.4. Defuzzifier

As we pointed out before, the main idea of a Mamdani FLS is to use crisp inputs to make fuzzy inference and finally find a crisp output which represents the behavior of the FLS. The process of finding a crisp output after fuzzification and inference is called *Deffuzification*. This final step consist on find an operation point given the results of the inference process of the FLS, which results on a fuzzy output set, so we need to use a mathematical method which returns a crisp measure of the behavior of the FLS.

There are many types of defuzzifiers, but we consider in this paper the *Height Defuzzifier* which replaces each rule output fuzzy set by a singleton at the point having maximum membership in that output set, $\bar{y}^l$, then it calculates the centroid of the resultantF set of these singletons. The crisp output of this defuzzifier is expressed as:

$$y(x) = f(x) = \frac{\sum_{l=1}^{M} \bar{y}^l \mu_{B^l}(\bar{y}^l)}{\sum_{l=1}^{M} \mu_{B^l}(\bar{y}^l)} \tag{9}$$

where $\bar{y}^l$ is the point having maximum membership in the output set [8].

2.3. Takagi-Sugeno-Kang (TSK) FLS

A TSK FLS is a special FLS which is also characterized by IF-THEN rules, but its consequent is a polynomial. Its output is a crisp value obtained from computing the polynomial output, so it does not need a defuzzification process. The l_{th} rule of a first order type-1 TSK FLS having p inputs $x_1 \in X_1, ..., x_p \in X_p$ and one output $y \in Y$ is expressed as:

$$R^l : \text{IF } x_1 \text{ is } F_1^l \text{ and } x_2 \text{ is } F_2^l \text{ and...and } x_p \text{ is } F_p^l$$

$$\text{THEN } y^l(x) = c_0^l + c_1^l x_1 + ... + c_p^l x_p \tag{10}$$

where $l = 1, ..., M$, $c_j^l (j = 0, .., p)$ are the consequent parameters, $y^l(x)$ is the output of the lth rule, and F_k^l $(k = 1, ..., p)$ are type-1 antecedent fuzzy sets.

The output of a TSK FLS is obtained by combining the outputs from the M rules in the following form:

$$y_{TSK}(x) = \frac{\sum_{l=1}^{M} f^l(x) \left(c_0^l + c_1^l x_1 + ... + c_p^l x_p \right)}{\sum_{l=1}^{M} f^l(x)} \tag{11}$$

where $f^l(x)$ $(l = 1, ..., M)$ are the rule firing levels and they are defined as:

$$f^l(x) = T_{k=1}^{p} \mu_{F_k^l}(x_k) \tag{12}$$

where T is a $t-norm$ operation, i.e. minimum or product operation (Mendel [8]), and x is the vector of inputs applied to the TSK FLS.

2.4. Fuzzy basis functions

For Mamdani FLSs, assuming that all consequent MFs are normalized, i.e., $\mu_{G^l}\left(\bar{y}^l\right) = 1$, and using singleton defuzzification, max-product composition and product implication, then the output of the height defuzzifier (9) becomes:

$$y(x) = f(x) = \frac{\sum_{l=1}^{M} \bar{y}^l T_{i=1}^{p} \mu_{F_i^l}(x_i)}{\sum_{l=1}^{M} T_{i=1}^{p} \mu_{F_i^l}(x_i)} \tag{13}$$

The FLS in (13) can be expressed as:

$$y(x) = f(x) = \sum_{l=1}^{M} \bar{y}^l \phi_l(x) \tag{14}$$

where $\phi_l(\mathbf{x})$ is called a *Fuzzy Basis Function* (FBF) of the *lth* rule [11], and it is defined as:

$$\phi_l(\mathbf{x}) = \frac{f^l}{\sum_{l=1}^{M} f^l} \qquad l = 1,\ldots,M \tag{15}$$

where f^l is given in (12).

This linear combination allows us to view an FLS as series expansions of FBFs [11], [1], [4] and [10] which has the capability of providing a mix of both numerical and linguistic information.

2.5. Weighted FBF

The crisp output of the TSK FLS in (11) can be expressed as:

$$y_{TSK}(\mathbf{x}) = \sum_{l=1}^{M} \phi_l(\mathbf{x}) \sum_{k=0}^{p} c_k^l x_k \tag{16}$$

It can also be expressed as:

$$y_{TSK}(\mathbf{x}) = \sum_{l=1}^{M} \sum_{k=0}^{p} \phi_k^l(\mathbf{x}) c_k^l \tag{17}$$

where $\phi_k^l(\mathbf{x})$ is the *kth* Weighted Fuzzy Basis Function (WFBF) of the *lth* rule which is expressed as [2]:

$$\phi_k^l(\mathbf{x}) = x_k \phi_l(\mathbf{x}), \quad l = 1,\ldots,M; k = 0,\ldots,p \tag{18}$$

This linear combination allows us to view the FLS as series expansions of WFBFs [2]. The WFBFs have also a capability of providing a combination of both numerical and linguistic information.

3. Orthogonality principle

We explain in this section how we can obtain, graphically, the optimal scalar that minimizes the norm of an error vector [9]. Suppose that we have a set of N measurements collected in a N-vector, $\vec{y}$, gathered for different values collected in another N-vector, $\vec{\phi}$. The problem is to find :

$$\min_{\theta} \left\| \vec{y} - \theta \vec{\phi} \right\| \tag{19}$$

As shown in Figure 2, we can see that the optimal scalar θ that minimizes the norm of the error vector, $\left\| \vec{e} = \vec{y} - \theta \vec{\phi} \right\|$, is obtained when $\vec{e} \perp \vec{\phi}$. This can be expressed as follows :

$$\vec{\phi} \cdot \left(\vec{y} - \theta \vec{\phi} \right) = 0 \tag{20}$$

Solving for θ we have:

$$\theta_{opt} = \frac{\vec{y}^T \vec{\phi}}{\vec{\phi}^T \vec{\phi}} \tag{21}$$

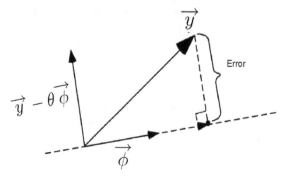

Figure 2. Basic Idea of Orthogonality Principle

4. FLS design based on GOP

GOP is an optimization principle which can be applied to both Mamdani and TSK FLSs. Under the premise of fixed shapes and the parameters of the antecedent MFs over the time, then a training dataset is used to tune the consequent parameters. The consequent parameters are c_k^l ($l = 1, ..., M$; $k = 0, ..., p$) in (11) for a TSK FLS, and $\bar{y}^l$ ($l = 1, ..., M$) in (9) for a Mamdani FLS.

4.1. Mamdani FLS design

Given a collection of N input-output numerical data training pairs

$$\left(\mathbf{x}^{(1)} : y^{(1)}\right), \left(\mathbf{x}^{(2)} : y^{(2)}\right), ..., \left(\mathbf{x}^{(N)} : y^{(N)}\right)$$

where $\mathbf{x}^{(i)}$ and $y^{(i)}$ are respectively the vector input and scalar output of the FLS given by (13). We have to tune the $\bar{y}^l$ ($l = 1, ..., M$) using these data training. Firstly, we compute the FBFs with training input vectors, then we apply the orthogonality principle on these FBFs and the training output vector.

Equation (14) can be decomposed as follows:

$$\begin{cases} y(\mathbf{x}^{(1)}) = f(\mathbf{x}^{(1)}) = \bar{y}^1\phi_1(\mathbf{x}^{(1)}) + ... + \bar{y}^M\phi_M(\mathbf{x}^{(1)}) \\ y(\mathbf{x}^{(2)}) = f(\mathbf{x}^{(2)}) = \bar{y}^1\phi_1(\mathbf{x}^{(2)}) + ... + \bar{y}^M\phi_M(\mathbf{x}^{(2)}) \\ \quad\quad\quad\quad\quad\quad \vdots \\ y(\mathbf{x}^{(N)}) = f(\mathbf{x}^{(N)}) = \bar{y}^1\phi_1(\mathbf{x}^{(N)}) + ... + \bar{y}^M\phi_M(\mathbf{x}^{(N)}) \end{cases} \quad (22)$$

So we have

$$y(\mathbf{x}^{(i)}) = f(\mathbf{x}^{(i)}) = \sum_{l=1}^{M} \bar{y}^l\phi_l(\mathbf{x}^{(i)}) \quad i = 1, ..., N \quad (23)$$

Now, if each FBF is considered as a basis function, we can compose the following vector:

$$\vec{\phi}_j = \begin{pmatrix} \phi_j(\mathbf{x}^{(1)}) \\ \phi_j(\mathbf{x}^{(2)}) \\ \vdots \\ \phi_j(\mathbf{x}^{(N)}) \end{pmatrix}, \quad j = 1, 2, ..., M \tag{24}$$

where M is the number of rules. We now collect all the N training output data in the same vector $\vec{y}$:

$$\vec{y} = \begin{pmatrix} y(\mathbf{x}^{(1)}) \\ y(\mathbf{x}^{(2)}) \\ \vdots \\ y(\mathbf{x}^{(N)}) \end{pmatrix} \tag{25}$$

and the parameters of the consequent in a vector $\vec{\theta}$:

$$\vec{\theta} = \begin{pmatrix} \bar{y}^1 \\ \bar{y}^2 \\ \vdots \\ \bar{y}^M \end{pmatrix} \tag{26}$$

By considering the N equations, a FLS can be expressed in vector-matrix format as follows:

$$\vec{y} = \Phi \vec{\theta} \tag{27}$$

where the fuzzy basis function matrix Φ is given by:

$$\Phi = [\vec{\phi}_1, \vec{\phi}_2, ..., \vec{\phi}_M] \tag{28}$$

To find the optimal vector $\vec{\theta}$ and because of fitting with basis sets, we generalize the presented orthogonality principle to a multi-dimensional basis leading to a GOP. The error vector should be perpendicular to all of the basis fuzzy vectors, as shown in Figure 2.

In a matrix form, we obtain:

$$\Phi^T \cdot \left(\vec{y} - \Phi \vec{\theta} \right) = 0 \tag{29}$$

Solving for $\vec{\theta}$, we have:

$$\vec{\theta}_{opt} = \begin{pmatrix} \bar{y}^1 \\ \bar{y}^2 \\ \vdots \\ \bar{y}^M \end{pmatrix} = \left[\Phi^T \Phi \right]^{-1} \Phi^T \vec{y}$$

where $\vec{\theta}_{opt}$ is a vector which contains the parameters of the consequent, i.e., $\bar{y}^l$ in (3).

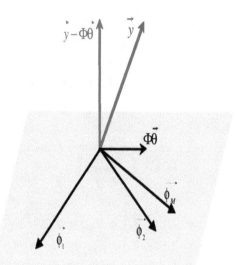

Figure 3. Basic Idea of Generalized Orthogonality Principle. The error vector should be perpendicular to all of the basis fuzzy vectors.

4.2. TSK FLS design

In the same way, the consequent parameters of a TSK FLS are tuned. The design approach is related to the following problem:

Given a collection of N input-output numerical training data pairs:

$$\left(\mathbf{x}^{(1)} : y^{(1)}\right), \left(\mathbf{x}^{(2)} : y^{(2)}\right), \dots, \left(\mathbf{x}^{(N)} : y^{(N)}\right)$$

where $\mathbf{x}^{(i)}$ is the $(p+1) - dimensional$ input vector ($p+1$ inputs with $x_0 \equiv 1$) and $y^{(i)}$ is the scalar output of the FLS given by (11). We have to tune the c_k^l ($l = 1, ..., M; k = 0, ..., p$) using these data training.

The WFBF vectors are computed using the training input data, then the GOP is applied to the $(p+1)$ combinations of WFBF vectors and the $(p+1)$ of $N-dimensional$ training output vector.

Using the elements of the input-output training pairs, the TSK output given in (17), can be rewritten as follows:

$$y_{TSK}(\mathbf{x}^{(i)}) = \begin{cases} \begin{bmatrix} \phi_0^1(\mathbf{x}^{(i)}) \\ \cdots \\ \phi_p^1(\mathbf{x}^{(i)}) \end{bmatrix}^T \begin{bmatrix} c_0^1 \\ \vdots \\ c_p^1 \end{bmatrix} \\ + \cdots + \\ \begin{bmatrix} \phi_0^M(\mathbf{x}^{(i)}) \\ \cdots \\ \phi_p^M(\mathbf{x}^{(i)}) \end{bmatrix}^T \begin{bmatrix} c_0^M \\ \vdots \\ c_p^M \end{bmatrix} \end{cases} \tag{30}$$

where $\mathbf{x}^{(i)} = \left[1, x_1^{(i)}, ..., x_p^{(i)}\right]^T$. Collecting the N equations we obtain:

$$\overrightarrow{y_{TSK}} = \begin{cases} \begin{bmatrix} \phi_0^1(\mathbf{x}^{(1)}) & \cdots & \phi_p^1(\mathbf{x}^{(1)}) \\ & \ddots & \\ \phi_0^1(\mathbf{x}^{(N)}) & \cdots & \phi_p^1(\mathbf{x}^{(N)}) \end{bmatrix} \begin{bmatrix} c_0^1 \\ \vdots \\ c_p^1 \end{bmatrix} \\ \qquad\qquad + \cdots + \\ \begin{bmatrix} \phi_0^M(\mathbf{x}^{(1)}) & \cdots & \phi_p^M(\mathbf{x}^{(1)}) \\ & \ddots & \\ \phi_0^M(\mathbf{x}^{(N)}) & \cdots & \phi_p^M(\mathbf{x}^{(N)}) \end{bmatrix} \begin{bmatrix} c_0^M \\ \vdots \\ c_p^M \end{bmatrix} \end{cases} \tag{31}$$

By taking each set of N WFBFs as a Weighted Fuzzy Basis Vector, WFBV:

$$\overrightarrow{\phi_k^l} = \begin{pmatrix} \phi_k^l(\mathbf{x}^{(1)}) \\ \phi_k^l(\mathbf{x}^{(2)}) \\ \vdots \\ \phi_k^l(\mathbf{x}^{(N)}) \end{pmatrix}, \quad \begin{cases} l = 1, ..., M \\ k = 0, ..., p \end{cases} \tag{32}$$

and each set of N outputs as a vector, the output vector can be expressed as follows :

$$\overrightarrow{y_{TSK}} = \begin{cases} \begin{bmatrix} \overrightarrow{\phi_0^1} & \cdots & \overrightarrow{\phi_p^1} \end{bmatrix} \begin{bmatrix} c_0^1 \\ \vdots \\ c_p^1 \end{bmatrix} \\ \qquad + \cdots + \\ \begin{bmatrix} \overrightarrow{\phi_0^M} & \cdots & \overrightarrow{\phi_p^M} \end{bmatrix} \begin{bmatrix} c_0^M \\ \vdots \\ c_p^M \end{bmatrix} \end{cases} \tag{33}$$

Now we have to tune $p+1$ parameters for each rule, i.e., M vectors of dimension $(p+1)$.

$$\overrightarrow{c^l} = \begin{pmatrix} c_0^l \\ \vdots \\ c_p^l \end{pmatrix}, \quad l = 1, ..., M \tag{34}$$

If we define the l_{th} element of Φ_{TSK} as $\Phi_{TSK,l}$, we have:

$$\Phi_{TSK,l} = \begin{bmatrix} \overrightarrow{\phi_0^l}, & \cdots, & \overrightarrow{\phi_p^l} \end{bmatrix}, l = 1, .., M \tag{35}$$

the output vector (33) becomes :

$$\overrightarrow{y_{TSK}} = \Phi_{TSK,1}\overrightarrow{c^1} + \cdots + \Phi_{TSK,M}\overrightarrow{c^M} \tag{36}$$

In a matrix form, (36) becomes :

$$\overrightarrow{y_{TSK}} = \Phi_{TSK} \left[\overrightarrow{c^1} \, \cdots \, \overrightarrow{c^M} \right]^T \tag{37}$$

So the *Weighted Basis Function Matrix* (WBFM) Φ can be defined as:

$$\Phi_{TSK} = \left[\Phi_{TSK,1}, \, \ldots, \, \Phi_{TSK,M} \right] \tag{38}$$

The optimal parameters of the consequent conforms a vector, $\overrightarrow{c^l}$ in (34) are obtained when the error vector, $\left(\overrightarrow{y_{TSK}} - \Phi_{TSK} \left[\overrightarrow{c^1} \, \cdots \, \overrightarrow{c^M} \right]^T \right)$, must be perpendicular to all the weighted fuzzy basis vectors, $\overrightarrow{\phi_k^l}$ $(k = 0, \ldots, p$ and $l = 1, \ldots, M)$, which are the columns of the WBFM Φ_{TSK}, as shown in Figure 4.

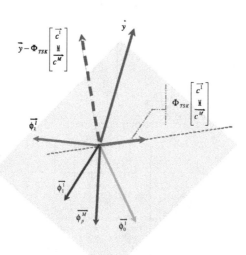

Figure 4. Extended Generalized Orthogonality Principle. The error vector $\overrightarrow{y} - \Phi \left[\overrightarrow{y_C^1} \, \cdots \, \overrightarrow{y_C^M} \right]^T$ should be perpendicular to all the fuzzy basis vectors, $\overrightarrow{\phi_k^l}$

This may be expressed directly in terms of the WBFM Φ as follows:

$$\Phi^T \overrightarrow{y} - \Phi^T \Phi \left[\overrightarrow{y_C^1} \, \cdots \, \overrightarrow{y_C^M} \right]^T = 0 \tag{39}$$

Solving for $\left[\overrightarrow{y_C^1} \, \cdots \, \overrightarrow{y_C^M} \right]^T$ provides the following

$$\left[\overrightarrow{y_C^1} \, \cdots \, \overrightarrow{y_C^M} \right]^T_{opt} = \left[\Phi \cdot \Phi^T \right]^{-1} \Phi \overrightarrow{y}$$

5. FLC design for controlling an inverted pendulum on a cart

5.1. Description of the system

Schematic drawing of an *Inverted pendulum On a Cart* (IPOC) system is depicted in Figure 5. where x is the position of the cart, θ is the angle of the pendulum with respect to the vertical direction and $\vec{F}$ is the external acting force in the $x - direction$. In order to keep the pendulum upright, we design a Fuzzy Logic Controller (FLC) using the GOP.

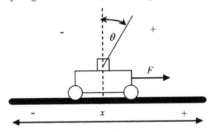

Figure 5. A schematic drawing of the inverted pendulum on a cart

The Lagrange equation for the position of the pendulum, θ, is given by:

$$\left(\frac{ml^2}{4} + J\right)\ddot{\theta} + \frac{ml}{2}(\ddot{x}\cos\theta - g\sin\theta) = 0 \tag{40}$$

The Lagrange equation for the position of the cart, x, is given by:

$$(M_1 + m)\ddot{x} + \frac{ml}{2}(\ddot{\theta}\cos\theta - \dot{\theta}^2\sin\theta) = F(t) \tag{41}$$

where J is the moment of inertia of the bar. The masses of the cart and the rod are $M_1 = 2Kg$ and $m = 0.1Kg$, respectively. The rod has a length $l = 0.5m$.

Since the goal of the control system is to keep the pendulum upright the equations can be linearized around $\theta = 0$. We chose $\mathbf{x} = \begin{bmatrix} \theta & \dot{\theta} & x & \dot{x} \end{bmatrix}^T$ as the state vector, where $\dot{\theta}$ is the pendulum angle variation and $\dot{x}$ is the cart position variation. The state representation is given by:

$$\dot{\mathbf{x}} = \begin{bmatrix} 0 & 1 & 0 & 0 \\ \frac{6}{l(m+4M_1)} & 0 & 0 & 0 \\ 0 & 0 & 0 & 1 \\ \frac{-3g \cdot m}{m+4M_1} & 0 & 0 & 0 \end{bmatrix} \mathbf{x} + \begin{bmatrix} 0 \\ \frac{6}{l(m+4M_1)} \\ 0 \\ \frac{4}{m+4M_1} \end{bmatrix} F(t) \tag{42}$$

$$\mu_{F_i^l}(x_i) = \exp\left[-\frac{1}{2}\left(\frac{x_i - m_{F_i^l}}{\sigma_{F_i^l}}\right)^2\right] \tag{43}$$

5.2. FLC structure and design

We try to keep the pendulum upright regardless the cart's position, i.e., *Pure Angular Position Control System* (PAPCS). Then, the two inputs of the Fuzzy Logic Controller FLC are the angular pendulum position, θ, and its derivative, $\dot{\theta}$, i.e., $x_1 = \begin{bmatrix} x_1 & x_2 \end{bmatrix} = \begin{bmatrix} \theta & \dot{\theta} \end{bmatrix}^T$ and its output is the applied force to the system $y = force$.

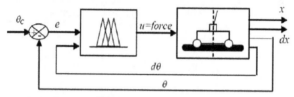

Figure 6. Fuzzy control system of the PAPCS

In this case, we use a Mamdani FLS with four rules. We use gaussian MF to fuzzify the two controller's inputs (44) and triangular MF to fuzzify the controller output.

$$\mu_{F_i^l}(x_i) = \exp\left[-\frac{1}{2}\left(\frac{x_i - m_{F_i^l}}{\sigma_{F_i^l}}\right)^2\right] \tag{44}$$

where $m_{F_i^l}$ and $\sigma_{F_i^l}$ are respectively the centers and standard deviations of these MFs.

The MFs of the antecedents are depicted in Figures 7 and 8.

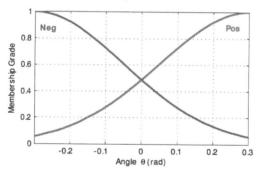

Figure 7. Membership functions for the first controller input θ

Figure 9 shows the 56 data training and the optimal fitting given by the GOP method.

The obtained optimal consequent parameters are

$$\left(\bar{y}^1, \bar{y}^2, \bar{y}^3, \bar{y}^4\right)_{opt} = (-14.3, -14.23, 9.61, 18.96)$$

Figure 10 shows the response of the pendulum system controlled by the designed FLC to a reference $\theta_{ref} = 0$ with its response at the same reference when it is controlled by untuned FLC. The initial state vector is $x_0 = \begin{bmatrix} \theta_0 & \dot{\theta}_0 & x_0 & \dot{x}_0 \end{bmatrix}^T = \begin{bmatrix} 0.1 & 0.2 & 0 & 0 \end{bmatrix}^T$.

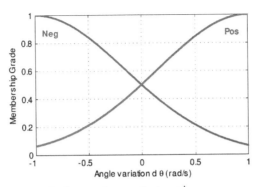

Figure 8. Membership functions for the second crontroller input $\dot{\theta}$

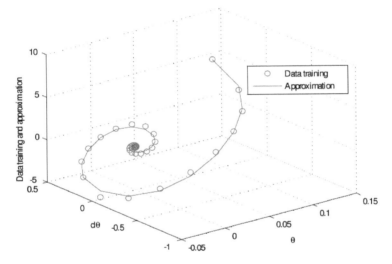

Figure 9. Data training and its approximation based on GOP

We evaluate the proposed design by using its error rate. For quantifying the errors, we use three different performance criteria to analyze the rise time, the oscillation behaviour and the behaviour at the end of transition period. These three criteria are: *Integral of Square Error* $(ISE = \int_0^\infty [e(t)]^2 \, dt)$, *Integral of the Absolute value of the Error* $(IAE = \int_0^\infty |e(t)| \, dt)$ and *Integral of the Time multiplied by Square Error* $(ITSE = \int_0^\infty t \, [e(t)]^2 \, dt)$

Table 1 summarizes the obtained values of *ISE*, *IAE* and *ITSE* of PAPCS using FLC, when tuning and no tuning are used.

We notice from this table that the errors obtained when tuning is used are all smaller than those obtained with untuned FLC. Fig. 11, 12, 13 show the different quantified errors.

Figures 10, 11, 12 and 13 show that the system using tuning is less oscillatory, having a rise time and errors at the end of transition period smaller than those obtained by untuned FLC.

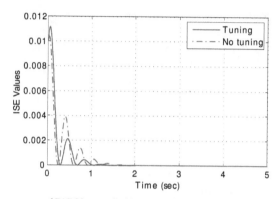

Figure 10. System responses of PAPCS controlled by a tuned and untuned FLC

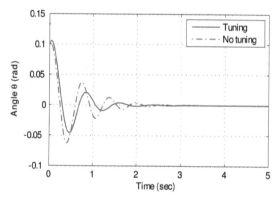

Figure 11. Integral of square error values of the PAPCS of tuned and no tuned consequent parameters. Rise time of the system is shorter for the tuned FLC

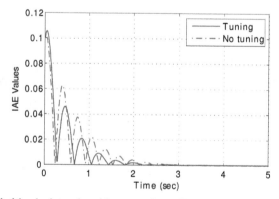

Figure 12. Integral of the absolute value of the error values of PAPCS of tuned and no tuned consequent parameters The system is less oscillatory for the tuned FLC before becoming stable

	No tuning	Tuning
ISE	0.2338	0.2224
IAE	4.7343	4.0403
ITSE	6.7733	4.4278

Table 1. Comparison of performance criteria for of PAPCS using tuned and no tuned FLC.

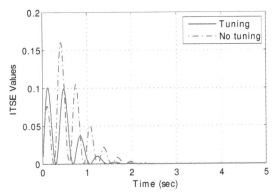

Figure 13. Integral of the time multiplied by square error values of PAPCS of tuned and no tuned consequent parameters. The error at the end of transition period is less important for the tuned FLC

6. FLS design for predicting time series

We apply the GOP to design an FLS which predicts a time series. The FLS has to predict the future value $x(t+6)$ of a Mackey-Glass time series (45) which is volatile. The following four antecedents were used: $x(t-18)$, $x(t-12)$, $x(t-6)$ and $x(t)$, which are known values of the time series ([2], [3]).

$$\frac{dx(t)}{dt} = \frac{0.2x(t-\tau)}{1+x^{10}(t-\tau)} - 0.1x(t) \tag{45}$$

The training data are obtained by simulating (45) for $\tau = 17$. We use the samples $x(1001), \cdots, x(1524)$ to train the IT2 FLS and the samples $x(1501), \cdots, x(2024)$ for testing. We use two Gaussian MFs per antecedent, so we have then 16 rules. The MFs of the antecedents are Gaussian, where its mean and the standard deviation were obtained from the 524 training samples, $x(1001), \cdots, x(1524)$. Table 2 summarizes the consequent parameters per each rule.

Figure 14 displays performance of the FLS in training data, and Figure 15 shows its results on Testing data. Note that the GOP-designed is a better forecaster, since the differences from original data are small in both training and testing data sets.

Some additional analyses should be performed to verify the goodness of fit of the method (See [5], and [6]), but in this case, the proposed GOP has shown good results, so we can recommend its application to real cases. Time series analysis is an useful topic for many decision makers, so the use of optimal and easy-to-be-implemented techniques, as the proposed one has a wide potential.

R^l	c_1	c_2	c_3	c_4	c_5
R^1	−3.58	7.94	−9.17	0.03	0.78
R^2	9.92	−10.9	−0.02	1.29	−7.83
R^3	16.05	7.58	12.40	5.25	−33.96
R^4	8.92	−6.78	−12.22	3.68	−4.13
R^5	1.06	4.64	28.42	−40.17	0.16
R^6	22.57	−33.28	−12.43	17.13	−12.76
R^7	−2.93	−7.65	5.73	−2.91	−1.30
R^8	22.88	26.23	−15.79	−6.19	−0.60
R^9	−3.86	4.36	2.04	0.21	4.77
R^{10}	27.72	−45.35	24.63	7.92	6.26
R^{11}	−0.24	−4.99	30.94	−26.54	6.65
R^{12}	2.36	5.34	−26.93	18.21	−8.03
R^{13}	−30.66	13.37	5.27	3.60	1.43
R^{14}	23.62	−21.97	−3.87	6.04	8.01
R^{15}	3.70	−5.07	0.61	−0.76	8.38
R^{16}	−25.05	11.30	−0.42	1.27	4.64

Table 2. The optimal TSK FLS consequent parameters obtained by GOP design.

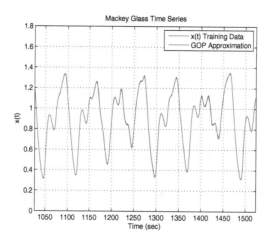

Figure 14. Mackey-Glass time series. The samples $x(1001), \cdots, x(1524)$ are used for designing the FLS forecaster

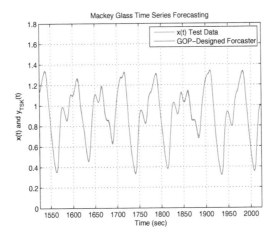

Figure 15. Output of the TSK FLS time-series forecaster. The samples $x(1525), \cdots, x(2024)$ are used for testing the GOP design

7. Concluding remarks

In this chapter we have presented an enhancement method of fuzzy controllers using the generalized orthogonality principle. We applied the method to two different cases: a first one involving control of an inverted pendulum and a second one for fuzzy forecasting. In the first application, numerical rules and their FBFs were extracted from numerical training data. This combination of both linguistic and numerical information simultaneously become FBFs an useful method. Since a specific FLS can be expressed as a linear combination of FBFs, we generalized orthogonality principle on FBFs that results in a better FLS.

In the second study case, we applied the GOP to design a FLS for time series forecasting. The FLS has been applied to a Mackey-Glass time series with better results compared to a non-GOP FLS. The results were validated with simulations.

All the FBFs can be seen as a basis vector, which allows to optimize the parameters of the consequents. This means that the error vectors are orthogonal to these FBFs, resulting in the minimization of the magnitudes of these error vectors, and consequently an optimal FLS.

The proposed method has a wide potential in complex forecasting problems ([5], and [6]). Its application to hardware design problems ([7]) can improve the performance of fuzzy controllers, so its implementation arises as a new field to be covered.

Author details

Nora Boumella
University of Batna, Batna - Algeria

Juan Carlos Figueroa
Universidad Distrital Francisco Jose de Caldas, Bogota - Colombia

Sohail Iqbal
NUST-SEECS, Islamabad - Pakistan

8. References

[1] Berenji, H. & Khedkar, P. [1992]. Learning and tuning fuzzy logic controllers through reinforcements, *IEEE Trans. Neural Networks* 3(1): 724–740.

[2] Boumella, N., Djouani, K. & Boulemden, M. [2011]. On an Interval Type-2 TSK FLS A1-C1 consequent parameters tuning, *in* IEEE (ed.), *Proc. SSCI 2011 T2FUZZ - 2011 IEEE Symposium on Advances in Type-2 Fuzzy Logic Systems*, IEEE, pp. 1–6.

[3] Boumella, N., Djouani, K. & Boulemden, M. [2012]. A robust Interval Type-2 TSK fuzzy logic system design based on chebyshev fitting, *International Journal of Control, Automation, and Systems* 10(4).

[4] Boumella, N., Djouani, K. & Iqbal, S. [2009]. A new design of fuzzy logic controller based on generalized orthogonality principle, *in* IEEE (ed.), *Proc. IEEE International Symposium on Computational Intelligence in Robotics and Automation, CIRA 2009*, IEEE, pp. 497–502.

[5] Figueroa, J. C. [2009]. An evolutive Interval Type-2 TSK fuzzy logic system for volatile time series identification, *2009 Conference on Systems, Man and Cybernetics*, IEEE, pp. 1–6.

[6] Figueroa, J. C., Kalenatic, D. & Lopez, C. A. [2010]. A neuro-evolutive Interval Type-2 TSK fuzzy system for volatile weather forecasting, *Lecture Notes in Computer Science* 6216: 142–149.

[7] Melgarejo, M. & Peña, C. A. [2007]. Implementing Interval Type-2 fuzzy processors I, *Computational Intelligence Magazine* 2(1): Pág. 63–71.

[8] Mendel, J. [2001]. *Uncertain Rule-Based Fuzzy Logic Systems: Introduction and New Directions*, Prentice Hall.

[9] Strang, G. [1988]. *Linear Algebra and its Applications, Third edition*, Brooks/Cole.

[10] Wang, L. [1992]. Fuzzy systems are universal approximators, *in* IEEE (ed.), *Proc. IEEE Int'l. Conf. on Fuzzy Systems, San Diego, CA*, IEEE, pp. 1163–1170.

[11] Wang, L. & Mendel, J. M. [1992]. Fuzzy basis functions, universal approximation, and orthogonal least squares learning, *IEEE Trans. Neural Networks* 3(5): 807–814.

Chapter 2

Hierarchical Fuzzy Control

Carlos André Guerra Fonseca, Fábio Meneghetti Ugulino de Araújo
and Marconi Câmara Rodrigues

Additional information is available at the end of the chapter

1. Introduction

Growing demands for comfort, reliability, accuracy, energy conservation, safety and economy have fueled interest in proposals that can contribute to facilitate high performance control systems design. In terms of vibrations active control, it may represent, for example, a good relationship between the maximum reduction in vibrations transmission between two systems and the minimum energy expended in order to accomplish this reduction [1].

The use of more than one controller to provide higher performance for complex systems has attracted interest because in each operation condition, their combination can take advantage of each controller's characteristics. To take advantage of controllers' combination, a supervisor can make a hierarchical classification of controllers' signals, according to the identified operational condition.

Advances in artificial intelligence, processing power and data storage, allowed the development of intelligent methods for different characteristics controllers' fusion. The use of intelligent methods allows to the controlled system: adaptability to various operational situations and proper performance, even in the presence of significant uncertainties. Intelligent supervisors are ease to maintain, to reconfigure and could have optimality during its operation according to the learning mechanism.

This chapter describes a methodology for controllers' combination called controllers hierarchical fusion. In this methodology, a supervisor system is used to obtain a single control signal from the control signals generated simultaneously by two or more controllers. A hierarchical controller's example compounded by one robust controller, one fuzzy controller and one fuzzy supervisor is applied for mechanical vibrations isolation and reference tracking using an electromechanical system proposed in [2]. This controller is called hierarchical fuzzy controller (HFC).

This electromechanical system can be used to eliminate vibrations in the camera of unmanned vehicles and also to position this camera. It can also be used in manned vehicles for drivers' seat positioning and to eliminate vibrations on it, as shown in Figure 1.

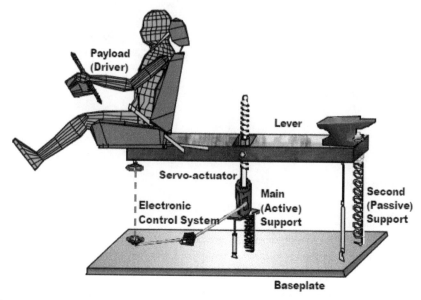

Figure 1. Application example: active suspension system

Digital simulations are employed in two case studies and the results are compared. On the first case study, the fuzzy controller and the fuzzy supervisor are tuned manually. Genetic algorithms (GA) are used on those systems tuning, in the second case study. Genetic algorithms usage facilitates designer's task and allows tuning parameters' optimization.

Next session describes the electromechanical system used and presents its models developed in [1]. The nonlinear model is used to validate the hierarchical fuzzy controller and in its fuzzy components' tuning, while the linearized model is used for robust control design. Performance criteria's are established at the end of this section.

2. Electromechanical system

Figure 2 details the electromechanical system used for vibration suppression and reference tracking. It consists on an l centimeters long bar with J inertia angular moment. It is considered that its mass m_B, is concentrated in its geometric center. This bar works as a lever which is supported in two points by systems with stiffness and damping, given by: k_A, k_B, c_A, c_B. In one extremity of the bar, a mass, m_A, called absorbing mass, is used to make a counterbalance with the payload. The payload is represented by a mass, m_C, on bar's free end. This system part is purely mechanical, being called lever system.

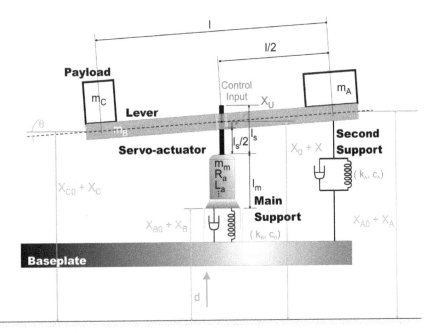

Figure 2. Electromechanical system

The vertical position control of bar's center is made by a servo actuator. This actuator consists of a DC servo motor whose axis is directly coupled to a spindle. The propeller's spindle step is given by L_P. It represents the direct relationship between motor's rotation angle (θ_M) and control's vertical displacement (X_u) imposed to bar's center with reference to the motor position (X_B).

The servo actuator varies the vertical position of bar's center depending on the measured displacements on bar's free end. This is done to isolate the payload from vibrations originated at the base.

A sensor that converts movements into voltage is used to measure vibrations on the payload. Those voltages feed servo motor, thus closing the control loop. Controllers are used to improve control efficiency, reaching thus performance specifications previously determined. This subsystem composed by one (or more) sensors, controllers and a servo-actuator, is called control system.

The nonlinear model used was developed in [1]. For the lever system it was given by:

$$\dot{x} = f\big(x(t), u(t), t\big)$$
$$y = g\big(x(t), t\big) = q_1 - \frac{l}{2} sen(q_2) \tag{1}$$

Where:

$$\mathbf{x}(t) = \begin{bmatrix} q_1 \\ \dot{q}_1 \\ q_2 \\ \dot{q}_2 \end{bmatrix}; \quad \mathbf{u}(t) = \begin{bmatrix} x_u \\ d \end{bmatrix}; \tag{2}$$

And:

$$\mathbf{f}(\mathbf{x}(t), \mathbf{u}(t), t) = \begin{bmatrix} \dot{q}_1 \\ \dfrac{-k_2 (T_{11} + T_{12}) + k_1 \cos(q_2)(T_{21} + T_{22})}{T_D} \\ \dot{q}_2 \\ \dfrac{-m(T_{21} + T_{22}) + k_1 \cos(q_2)(T_{11} + T_{12})}{T_D} \end{bmatrix} \tag{3}$$

With:

$$k_1 = \frac{l}{2}(m_A - m_C) \tag{4}$$

$$k_2 = \left(\frac{l}{2}\right)^2 (m_A + m_C) + \frac{1}{12} m_B (a^2 + l^2) \tag{5}$$

$$T_{11} = (k_A \delta_A + k_B \delta_B - mg) - \frac{1}{8} k_A (8q_1 + 4l\, sen(q_2) - 8d)$$
$$- lk_B (q_1 - x_U - d) - c_B (\dot{q}_1 - \dot{x}_U - \dot{d}) + k_1 \dot{q}_2^2 sen(q_2) + m_m \overline{x}_u \tag{6}$$

$$T_{12} = -\dfrac{\frac{1}{16} c_A (8q_1 + 4l\, sen(q_2) - 8d)}{4\left(q_1 + \frac{l}{2} sen(q_2) - d\right)^2 + l^2 (1 - \cos(q_2))^2}$$
$$\left[8\left(\dot{q}_1 + \frac{l}{2}\dot{q}_2 \cos(q_2) - \dot{d}\right)\left(q_1 + \frac{l}{2} sen(q_2) - d\right) + 2l^2 \dot{q}_2 sen(q_2)(1 - \cos(q_2))\right] \tag{7}$$

$$T_{21} = \frac{l}{2}(m_A - m_C)g - \frac{1}{8} k_A \left[4\left(q_1 + \frac{l}{2} sen(q_2) - d\right) l\cos(q_2) + 2l^2 (1 - \cos(q_2)) sen(q_2)\right] \tag{8}$$

$$T_{22} = -\dfrac{\frac{1}{16} c_A \left[4\left(q_1 + \frac{l}{2} sen(q_2) - d\right) l\cos(q_2) + 2l^2 (1 - \cos(q_2)) sen(q_2)\right]}{4\left(q_1 + \frac{l}{2} sen(q_2) - d\right)^2 + l^2 (1 - \cos(q_2))^2}$$
$$\left[8\left(\dot{q}_1 + \frac{l}{2}\dot{q}_2 \cos(q_2) - \dot{d}\right)\left(q_1 + \frac{l}{2} sen(q_2) - d\right) + 2l^2 \dot{q}_2 sen(q_2)(1 - \cos(q_2))\right] \tag{9}$$

$$T_D = -k_2 m + \left(k_1 \cos\left(q_2\right)\right)^2 \tag{10}$$

The equation that describes servo actuator dynamics is given by:

$$\ddot{x}_u + \frac{1}{T_m}\dot{x}_u = \frac{L_p K_m}{T_m} e_a \tag{11}$$

For robust control project it was used the linearized model founded in [1].

$$
\begin{bmatrix} \dot{x}_1 \\ \dot{x}_2 \\ \dot{x}_3 \\ \dot{x}_4 \\ \dot{x}_5 \\ \dot{x}_6 \end{bmatrix} =
\begin{bmatrix}
0 & 1 & 0 & 0 & 0 & 0 \\
-a_1^{12} & -a_1^{11} & -a_2^{12} & -a_2^{11} & -a_3^{12} & -a_3^{11} \\
0 & 0 & 0 & 1 & 0 & 0 \\
-a_1^{22} & -a_1^{21} & -a_2^{22} & -a_2^{21} & -a_3^{22} & -a_3^{21} \\
0 & 0 & 0 & 0 & 0 & 1 \\
0 & 0 & 0 & 0 & 0 & -a_3^{31}
\end{bmatrix}
\begin{bmatrix} x_1 \\ x_2 \\ x_3 \\ x_4 \\ x_5 \\ x_6 \end{bmatrix} +
\begin{bmatrix}
0 & \beta_2^{11} \\
\beta_1^{12} & \beta_2^{12} \\
0 & \beta_2^{21} \\
\beta_1^{22} & \beta_2^{22} \\
0 & 0 \\
\beta_1^{32} & 0
\end{bmatrix}
\begin{bmatrix} e_a \\ d \end{bmatrix}
$$

$$
y = x_c = \begin{bmatrix} 1 & 0 & -\dfrac{l}{2} & 0 & 0 & 0 \end{bmatrix} \cdot
\begin{bmatrix} x_1 \\ x_2 \\ x_3 \\ x_4 \\ x_5 \\ x_6 \end{bmatrix}
\tag{12}
$$

The system states are:

$$x_1 = q_1 = x,\ x_2 = \dot{q}_1 - \beta_2^{11} d = \dot{x} - \beta_2^{11} d,\ x_3 = q_2 = \theta,\ x_4 = \dot{q}_2 - \beta_2^{21} d = \dot{\theta} - \beta_2^{21} d,\ x_5 = \theta_m,\ x_6 = \dot{\theta}_m \tag{13}$$

Where:

$$\beta_1^{12} = b_1^{12},\ \beta_1^{22} = b_1^{22},\ \beta_1^{32} = b_1^{32},\ \beta_2^{11} = b_2^{11},\ \beta_2^{21} = b_2^{21},\ \beta_2^{12} = b_2^{12} - a_1^{11} b_2^{11} - a_2^{11} b_2^{21},$$
$$\beta_2^{22} = b_2^{22} - a_1^{21} b_2^{11} - a_2^{21} b_2^{21} \tag{14}$$

The coefficients a_i^{jk} and b_i^{jk} are given by:

$$a_1^{11} = -CL_{11},\ a_1^{12} = -CL_{12},\ a_2^{11} = -CL_{13},\ a_2^{12} = -CL_{14},\ a_3^{11} = -L_p CL_{18} + \frac{L_p CL_{17}}{T_m},$$
$$a_3^{12} = -L_p CL_{19},\ a_1^{21} = -CL_{21},\ a_1^{22} = -CL_{22},\ a_2^{21} = -CL_{23},\ a_2^{22} = -CL_{24}, \tag{15}$$
$$a_3^{21} = -L_p CL_{28} + \frac{L_p CL_{27}}{T_m},\ a_3^{22} = -L_p CL_{29},\ a_3^{31} = \frac{1}{T_m}$$

And:

$$b_1^{12} = \frac{L_p K_m CL_{17}}{T_m}, \; b_2^{11} = CL_{15}, \; b_2^{12} = CL_{16}, \; b_1^{22} = \frac{L_p K_m CL_{27}}{T_m}, \; b_2^{21} = CL_{25}, \; b_2^{22} = CL_{26}$$

$$b_1^{32} = \frac{K_m}{T_m}$$

(16)

Where:

$$CL_{11} = \frac{-k_2\left(-c_a - c_b\right) - \frac{l}{2}k_1 c_a}{-k_2 m + k_1^2}$$

(17)

$$CL_{12} = \frac{-k_2\left(-k_a - lk_b\right) - \frac{l}{2}k_1 k_a}{-k_2 m + k_1^2}$$

(18)

$$CL_{13} = \frac{\frac{l}{2}k_2 c_a - \left(\frac{l}{2}\right)^2 k_1 c_a}{-k_2 m + k_1^2}$$

(19)

$$CL_{14} = \frac{\frac{l}{2}k_2 k_a - \left(\frac{l}{2}\right)^2 k_1 k_a}{-k_2 m + k_1^2}$$

(20)

$$CL_{15} = \frac{-k_2\left(c_a + c_b\right) + \frac{l}{2}k_1 c_a}{-k_2 m + k_1^2}$$

(21)

$$CL_{16} = \frac{-k_2\left(k_a + lk_b\right) + \frac{l}{2}k_1 k_a}{-k_2 m + k_1^2}$$

(22)

$$CL_{17} = \frac{-k_2 m_m}{-k_2 m + k_1^2}$$

(23)

$$CL_{18} = \frac{-k_2 c_b}{-k_2 m + k_1^2}$$

(24)

$$CL_{19} = \frac{-k_2 lk_b}{-k_2 m + k_1^2}$$

(25)

$$CL_{21} = \frac{\frac{l}{2}mc_a + k_1\left(-c_a - c_b\right)}{-k_2 m + k_1^2}$$

(26)

$$CL_{22} = \frac{\frac{l}{2}mk_a + k_1\left(-k_a - lk_b\right)}{-k_2m + k_1^2}$$ (27)

$$CL_{23} = \frac{\left(\frac{l}{2}\right)^2 mc_a - \frac{l}{2}k_1c_a}{-k_2m + k_1^2}$$ (28)

$$CL_{24} = \frac{\left(\frac{l}{2}\right)^2 mk_a - \frac{l}{2}k_1k_a}{-k_2m + k_1^2}$$ (29)

$$CL_{25} = \frac{-\frac{l}{2}mc_a + k_1\left(c_a + c_b\right)}{-k_2m + k_1^2}$$ (30)

$$CL_{26} = \frac{-\frac{l}{2}mk_a + k_1\left(k_a + lk_b\right)}{-k_2m + k_1^2}$$ (31)

$$CL_{27} = \frac{k_1 m_m}{-k_2m + k_1^2}$$ (32)

$$CL_{28} = \frac{k_1 c_b}{-k_2m + k_1^2}$$ (33)

$$CL_{29} = \frac{k_1 lk_b}{-k_2m + k_1^2}$$ (34)

Nonlinear system response to a step reference and for a step disturb was used to determine the performance criteria.

Figure 3 shows the nonlinear system in closed loop, without controllers, step response. This response is characterized by the influence of two vibrations modes: one slower and overdamped and the other faster and oscillating. It practically has no overshoot. The settling time, considering an accommodation range of ± 5% of the reference signal amplitude, is more than 12.5s. The rise time from 0 to100% of the reference signal amplitude is greater than 19s. This large difference between the rise time and the settling time highlights the influence of the overdamped mode [3].

Figure 4 shows the non-controlled system response to a disturbance.

With the reference fixed at zero, when a 0.01m amplitude step disturbance is injected into the system without the controller, its output goes upper than one and a half the amplitude of the injected disturbance. The non-controlled system needs about 12.8s to reject this disturbance on the mentioned condition, considering that the disturbance is sufficiently rejected when the response amplitude is reduced to a range of ± 5% of the injected disturbance amplitude, around zero. Figure 4 shows this response.

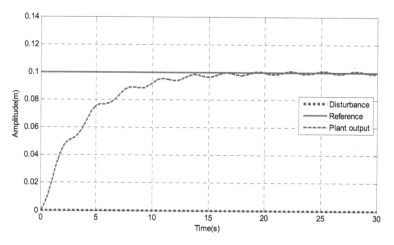

Figure 3. Electromechanical system step response without controllers and disturbance

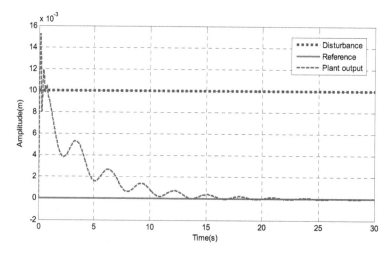

Figure 4. Non-controlled system response to a step disturbance

Thus, the performance specifications that characterize a satisfactory response to the nonlinear system are: A step reference signal must be tracked without regime error; the rise time should be reduced to at most 10% of the time obtained by the non-controlled system;

The settling time should be reduced to at most 20% of the time obtained with the non-controlled system; The overshoot should be less than 10%; The time required for the controlled system to reject a step disturbance, must be reduced by at least 95%; Furthermore, the response signal may not exceed 40% of disturbance's amplitude; Finally, the control signal generated must respect the servo-actuator saturation limits, that, in this case, is ± 15V.

Those specifications were achieved through the use of the hierarchical fuzzy controller. Each controller design aimed to meet some performance specifications. In that way, conflicting specifications were separately addressed, instead of trying, in each project, to get a fit to satisfy conflicting specifications, relaxing those specifications. So the hierarchical fuzzy controller should take the best features of each controller, to meet all the specifications described in this section.

3. Robust control

In vibration control, as well as in several other applications, it is desired that the control system presents robustness to the effects of factors such as: modeling errors, variations in the parameters of the system being controlled, noise and disturbances. There are at least two reasons why the robustness is a desirable feature in the control systems: the need of control systems that operates satisfactorily, even in operating conditions different from the ones considered in the model design; and the possibility to adopt an intentionally simplified project model, to reduce: the time spent in the modeling stage and the resulting controller complexity [4].

Among the main techniques for robust controllers synthesis can be cited: The Linear Quadratic Gaussian / Loop Transfer Recovery (LQG/LTR), H2 and H∞ optimizations, methods based on Lyapunov functions, minmax optimization and Quantitative Feedback Theory (QFT).

The LQG/LTR controller designed in [1] was used to allow a better comparison between the optimized hierarchical fuzzy controller implemented and the non-optimized developed in [1]. Furthermore the LQG/LTR technique has a simple and systematic design procedure, the controller robustness is ensured by this procedure, even in a broad class modeling errors presence and also the number of design parameters is relatively small [5].

This procedure has two steps: initially the target filter loop (TFL) must be projected. It must meet the performance specifications previously established. Once obtained an appropriate TFL, its characteristics are recovered for the transfer function of the loop formed by the controller and the nominal model $\left(G_K(s) \cdot G_N(s)\right)$.

The LTR procedure, initially proposed in [6], suggests that the TFL is achieved through the design of a Linear Quadratic Regulator (LQR) and then recovered by adjusting a Kalman filter. Another way to do it is to set a Kalman filter, to obtain a satisfactory target filter loop, and then project an optimal state feedback, type LQR, to recover the TFL [1].

Given the linearized model in form:

$$\dot{\mathbf{x}}(t) = \mathbf{A}\mathbf{x}(t) + \mathbf{B}\mathbf{u}(t)$$
$$\mathbf{Y}(t) = \mathbf{C}\mathbf{x}(t) \tag{35}$$

The Kalman's filter design begins with the solution of the following algebraic Riccati equation:

$$\mathbf{A\Sigma} + \mathbf{\Sigma A}^{T} + \mathbf{W\Xi W}^{T} - \mathbf{\Sigma C}^{T}\mathbf{\Theta}^{-1}\mathbf{C\Sigma} = 0 \tag{36}$$

In [1] it was used:

$$\mathbf{W} = \mathbf{B}(:,1); \mathbf{\Xi} = \mathbf{I}; \mathbf{\Theta} = \mu\mathbf{I} \tag{37}$$

Where $\mathbf{B}(:,1)$ corresponds to the first column of the B matrix and μ is the project's free parameter. This choice was made because the first attempt to select the W matrix must be the matrix related with the control input [5]. As could be seen in [1], this choice proved satisfactory.

In [1] were also used: $\mu = 10^{-6}$ to obtain the TFL and $\rho = 10^{-12}$ to recover the TFL, resulting in a LQG/LTR robust controller with the following desired characteristics: good speed in test model controlled response accommodation, when tracking a reference, and principally a good rejection of disturbances. Figure 5 illustrates the TFL obtained and recovered for these values of μ and ρ.

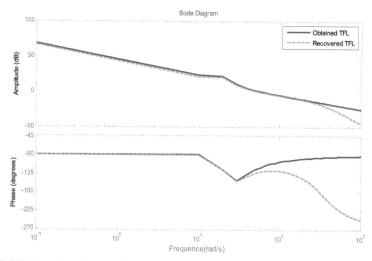

Figure 5. TFL obtained and recovered

As mentioned earlier, this LQG/LTR controller was used to allow a better comparison between the optimized hierarchical fuzzy controller implemented and the non-optimized developed in [1].

With the system controlled only by this robust controller, a step reference with 0.1m amplitude, was tracked without regime error; the rise time from 0 to 100% of the reference, in disturbance absence, was about 0.03s which corresponds to 0.16% of the rise time obtained by the non-controlled system; the settling time for (± 5%) was 0.17s, so, it was reduced to 1.36% of the time obtained with the non-controlled system; the overshoot was 22.4% and the control signal generated to track this reference signal, surpassed the actuator saturation levels. Therefore, with respect to the reference tracking, the controller could not satisfy two performance criteria established, because the overshoot was higher than 10% of the reference signal and some control signals produced, extrapolates the servo actuator saturation levels. Figures 6 and 7 show the system response when controlled only by this LQG/LTR robust controller.

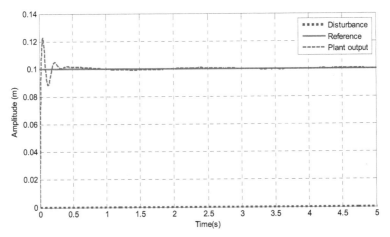

Figure 6. System response on step reference tracking, only with the robust controller, and in disturbances absence

With a null reference, a 0,01m step disturbance was injected in the system. The time required for the system to reject this disturbance using only the robust controller, was approximately 0.17s; what represents a 98.67% time reduction when compared to non-controlled system exposed to the same situation; The response signal maximum amplitude was 17.89% of the disturbance amplitude; the control signal varied within the levels of the servo actuator saturation. So in disturbance rejection, with null reference, the robust controller met all performance requirements described, as could be seen on figures 8 and 9.

It was also evaluated the system response, only with the robust controller, to a square wave reference with 0.1m peak to peak, 0.015Hz frequency and 100s duration. The system tracked this reference without regime error, the rise time and the settling time satisfied the performance specifications, but, again, as was expected, the control signal exceeded the actuator saturation limits and the overshoot exceeded the maximum stated in performance criteria, as could be seen on figures 10 and 11.

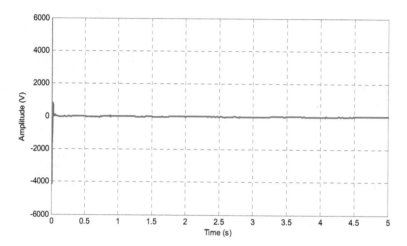

Figure 7. Robust controller signal for a step reference tracking, in disturbances absence

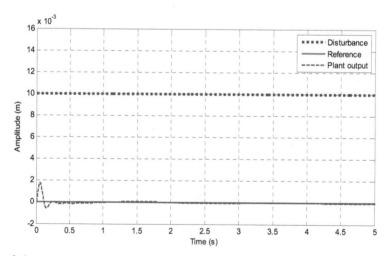

Figure 8. System response on step disturbance rejection, only with the robust controller, and with a null reference

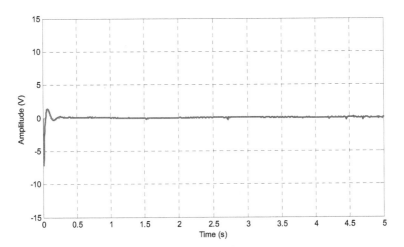

Figure 9. Robust controller signal for a step disturbance rejection, with a null reference

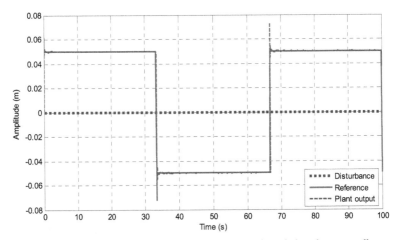

Figure 10. System response on square wave reference tracking, only with the robust controller

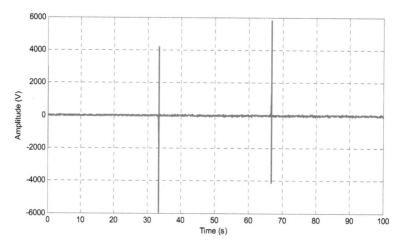

Figure 11. Robust controller signal for a square wave reference tracking, in disturbances absence

Finally, the system, only with the robust controller, was tested on tracking a step reference in the presence of uniformly distributed white noise with 0.02m peak to peak. Figures 12 and 13 show the system response and the control signal applied to the plant in this situation.

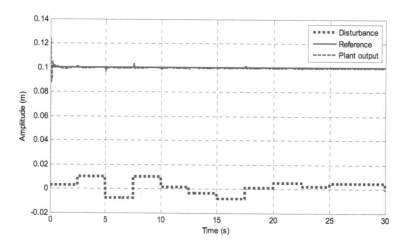

Figure 12. System response on step reference tracking, using only the robust controller, and in uniformly distributed white noise presence

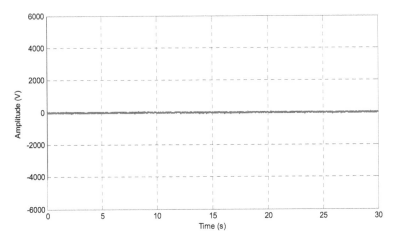

Figure 13. Robust controller signal for a step reference tracking, in uniformly distributed white noise presence

Comparing those results with the first shown, it is concluded that the rise time and the settling time were the same for both situations. In the white noise presence, the system showed a slightly higher overshoot, 23.4%, what is unsatisfactory according to the performance criteria, as well as the control signal applied that extrapolates the actuator saturation limits. So, as expected, in both cases the same performance requirements were not satisfied.

Therefore, those requirements should be met by the fuzzy controller and the supervisor must properly combine those two controllers to meet all performance criteria.

4. Fuzzy control

Fuzzy controllers are those that make use of fuzzy logic, which is based on the fuzzy sets' theory. This theory was developed by Zadeh in 1965 [7], to deal with the vague aspect of information through the mathematical representation of expressions commonly used by humans, also called linguistic variables, which give a not exact value to a variable characteristic of the object under observation.

Fuzzy logic attaches to a statement, not the value 'true' or 'false', but a veracity degree within a numeric range.

Due to its ability to handle uncertainty and imprecision, fuzzy logic has been characterized as one of the current technologies for the successful development of systems to control sophisticated processes, enabling the use of simple controllers to satisfy complex design requirements, even when the model of the system to be controlled has uncertainties [8-14].

The greatest difficulty in creating fuzzy systems is the definition of linguistic terms and rules. One way to solve this problem is to use hybrid approaches as models called neuro-fuzzy. In a neuro-fuzzy system those parameters are learned with the presentation of training pairs (input, desired output) to a neural network whose nodes basically computes intersection and union operators [15-18]. Another hybrid approach that allows the parameters tuning for fuzzy systems, consists in the use of genetic algorithms [19].

A satisfactory definition of the number of membership functions and the degree of overlap between them is fundamental when implementing a fuzzy controller. It directly influences on the next stage, called inference [20].

The inference uses a set of rules that describe the dependence between the linguistic variables of input and output functions. This relationship is usually determined heuristically and consists of two steps: aggregation, when evaluating the 'if' part of each rule, through the operator "and fuzzy," and the composition stage, using the operator "or fuzzy" to considering the different conclusions of the active rules [20, 21].

After the inference from the action to be taken, the classical fuzzy models require a decoding of the linguistic value for the numeric variable output, called defuzzification. This output can represent functions such as adjusting the position of a button, or provide voltage to a particular motor.

The Takagi Sugeno fuzzy controllers do not need a defuzzification step, because they obtain this precise equivalence directly [9, 19]. Therefore they were used to compound the fuzzy hierarchical controller.

For the design and optimization of the fuzzy logic controller it was used the nonlinear model of the physical system, as this model provides a more accurate representation of it.

All available knowledge about the system being controlled is of fundamental importance for the initial stage of designing a fuzzy controller, therefore, knowing the geometrical characteristics, the dynamics and any system particularity, can significantly reduce the project effort [1]. The fuzzy logic controller used has the following structure: Two inputs, which are: the tracking error (the difference between the reference and the system output) and its derivative; an output which is the control signal. For the output variable composition 25 first-order Sugeno functions are used; five linguistic variables were defined for each input variable: Negative Big, Negative Small, Zero, Positive Small and Positive Big; Triangular membership functions were chosen for the input variables; The probabilistic t-norm and t-conorm operators were chosen; The rule base is composed by 25 rules. For each rule there is a Sugeno output function; For the inference procedure, the Sugeno interpolation model was chosen.

The tuning of this fuzzy controller was made by a genetic algorithm. This algorithm is based on the laws of natural selection and evolution. It searches to an optimal solution in the space of solutions given by the designer, using probabilistic rules for combining solutions in order to improve their quality. It is therefore an efficient search strategy that can be used in optimization or classification problems [22-25].

In the fuzzy controller's optimization, each individual is formed by 70 genes. The first 20 genes represent the input membership functions. The 50 subsequent genes describe the coefficients of the Sugeno output functions, t_i and s_i. Those functions are given by:

$$\left[t_i \cdot e + s_i \cdot \frac{de}{dt} + 0 \right]$$ (38)

Where: e is the error.

With the use of genetic algorithms for tuning of all parameters of fuzzy controller, the designer's task is to limit the search space of GA and find a good setting of its parameters, in order to obtain the desired results.

The determination of the limits of the search spaces for the fuzzy controller optimization was based on the results obtained in [1] and in several tests. The population size, the percentage of mutation and the stopping criteria were also determined from several tests.

To obtain the results that will be shown, a square wave was used as reference, allowing a good fit to the fuzzy controller for several references. The genetic algorithm configuration was: population of 30 individuals, all children were generated by recombination with mutation probability of 5% for each gene; the roulette method was used on selection step. The stopping criteria were: maximum number of iterations equal to 100, repeating the best individual for 25% of the generations' maximum number, maintaining the average fitness of the population for 10% of the generations' maximum number and mean square error of 10^{-5}.

For the evaluation of each individual the control of the nonlinear system using only the fuzzy controller, was simulated during 100s. The evaluation function used for this controller tuning, was:

$$f_{ev}(ind.) = 8t_{r1} + 8t_{r2} + 8t_{r3} + 0,9o_{s1} + 0,9|o_{s2}| + 0,9o_{s3} + 0,7t_{s1} + 0,7t_{s2}$$
$$+ 0,7t_{s3} + 15e_m^2 + u_{max} - u_{min}$$ (39)

Where: the "$t_{ri's}$" are the rise times, the "$o_{si's}$" are the overshoots, the "$t_{si's}$" are the settling times, "e_m" is the average error, "u_{max}" is the maximum positive amplitude of the control signal above actuator's saturation and "u_{min}" is the maximum amplitude of the negative control signal, below actuator's saturation.

Higher weights were given to the mean square error and to the rise times because it was observed that they had a lower representation in the evaluation function, than the settling time and the peaks of the control signals above actuator's saturation. Thus allowing to the genetic algorithm, the search for a tune that provides not only short settling times through low control signals, but also small rise times, and that the system does not presents regime errors. Lower weights were given for the settling times and the overshoots, to allow the search for fuzzy controllers that give the system a higher speed.

Figures 14 and 15, shows the fuzzy controller optimized membership functions.

Two search spaces were defined for output functions' coefficients determination: one from 0 to 100, for the coefficients of the functions associated with rules that involve in its antecedent the linguistic variables negative big or positive big, and another from 0 to 60 to the coefficients of the other functions. The independent terms of output functions were not optimized and were always made equal to zero. The output functions obtained after the tuning can be seen in Table 1.

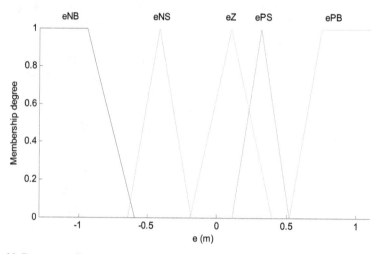

Figure 14. Fuzzy controller optimized membership functions of error input

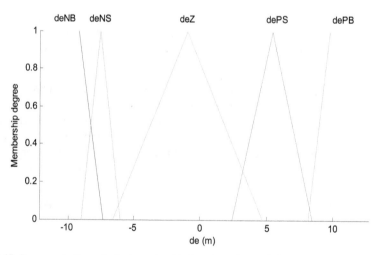

Figure 15. Fuzzy controller optimized membership functions of error derivative input

Function name	Parameters [t s]
S1	[10.93 94.07]
S2	[95.34 3.74]
S3	[25.50 27.17]
S4	[39.12 62.36]
S5	[95.86 94.12]
S6	[28.08 47.32]
S7	[46.76 43.30]
S8	[27.71 13.14]
S9	[42.55 5.52]
S10	[51.22 41.09]
S11	[53.64 5.98]
S12	[31.55 12.47]
S13	[53.31 1.20]
S14	[37.24 16.27]
S15	[33.20 46.69]
S16	[21.36 29.04]
S17	[2.97 23.94]
S18	[11.98 43.12]
S19	[42.94 45.81]
S20	[10.77 27.04]
S21	[75.33 50.81]
S22	[85.35 66.19]
S23	[45.64 50.41]
S24	[58.09 15.50]
S25	[15.18 11.76]

Table 1. Output functions' parameters of the optimized fuzzy controller

Table 2 shows the fuzzy controller rule base.

The control of the electromechanical system made only by the optimized fuzzy controller, presented a poor performance in tracking a 0.1m amplitude step reference, in disturbance absence. The overshoot presented was out of performance specifications (30.60%), and the settling time was almost equal to the uncontrolled system settling time (11.09s). However, the system showed no error at steady state, the rise time was satisfactory, 0.22 s, and the control signal produced was far below the actuator saturation, allowing the use of this controller in the hierarchical control scheme, as a supplier of control signals applicable in situations of great error, where the signals produced by the robust controller extrapolate the servo-actuator saturation. Those results are shown in figures 16 and 17.

		Error				
		eNB	eNS	eZ	ePS	ePB
Error derivative	deNB	S1	S6	S11	S16	S21
	deNS	S2	S7	S12	S17	S22
	deZ	S3	S8	S13	S18	S23
	dePS	S4	S9	S14	S19	S24
	dePB	S5	S10	S15	S20	S25

Table 2. Rule base of fuzzy controller

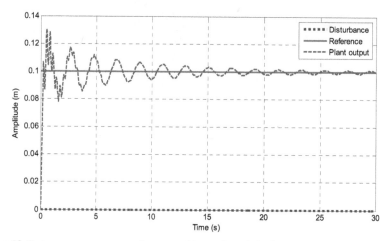

Figure 16. System response on step reference tracking, only with the fuzzy controller, and in disturbances absence

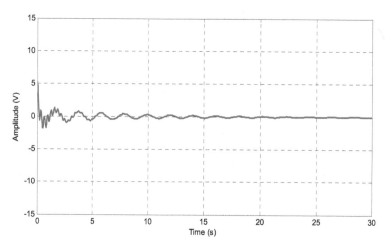

Figure 17. Fuzzy controller signal for a step reference tracking, in disturbances absence

In the rejection of a 0.01m amplitude step disturbance with the null reference, the system response with the fuzzy controller was also unsatisfactory, because its amplitude exceeded in 21% the disturbance amplitude, and it took about 2.01s to reject it, far above the 0.64s, established as a goal. Figures 18 and 19 show the system response and the control signal for this case.

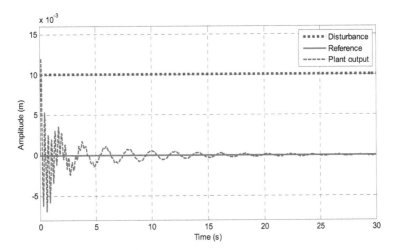

Figure 18. System response on step disturbance rejection, only with the fuzzy controller, and with a null reference

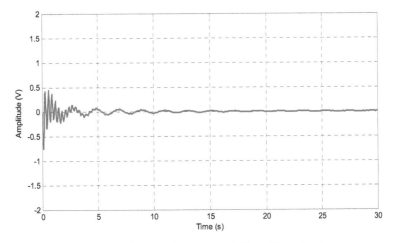

Figure 19. Fuzzy controller signal for a step disturbance rejection, with a null reference

It was also evaluated the system response on a square wave reference tracking in the absence of disturbances and using only the fuzzy controller. As can be seen in figures 20 and 21 the system tracked the reference without regime error, the rise times were acceptable, but the settling times were greater than desirable, moreover, the overshoot and the control signal extrapolated performance specifications. But the fuzzy controller's peak signal was much lower than the robust one.

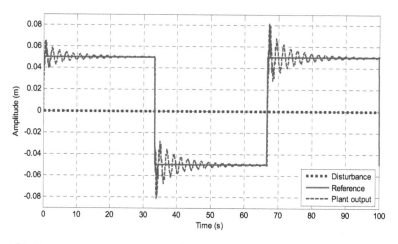

Figure 20. System response on square wave reference tracking, only with the fuzzy controller

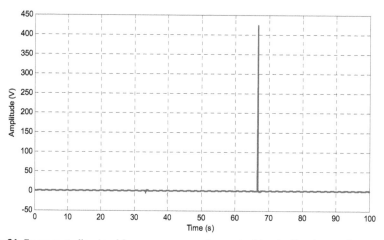

Figure 21. Fuzzy controller signal for a square wave reference tracking, in disturbances absence

From these results, it can be concluded that the function of the fuzzy controller is to bring the plant to a situation that favors the use of the robust controller, avoiding the extrapolation of control signal limits.

5. Fuzzy supervisor

The multiple controllers' fusion seeks to achieve higher performance than those obtained using only one controller.

The supervisor's task is to find an ideal combination of control signals generated by the controllers designed, in such way that this combination compose the control signal which will effectively act on the plant. To do this, the supervisor evaluates the operating condition in each instant, and then determines an importance hierarchy of each control signal. Therefore, in addition to control signals generated by the controllers, the supervisor must also receive information that enables to evaluate the operating condition at all instants, and then, based on this evaluation, the supervisor will sort, hierarchically, the outputs of the controllers, compounding then the control signal that will act on the plant. This hierarchy is the level of importance associated by the supervisor to each controller in every operating condition. It defines the participation of each controller in the control signal that will be applied on the plant.

The fuzzy supervisor used was a Takagi-Sugeno system with: two inputs, which are the same used in the fuzzy controller; 3 linguistic variables (negative, zero and positive), which are represented by trapezoidal membership functions; two output functions, which are zero order functions.

Figure 22 illustrates the architecture used for the control signals fusion via hierarchical fuzzy supervisor.

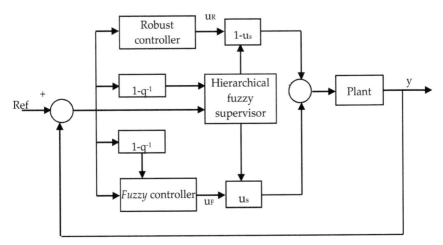

Figure 22. Control scheme using the fuzzy hierarchical controller

From the difference between a reference signal, specified by the operator, and the vertical position of the bar's free end, measured by a sensor, it is produced an error signal. With this error signal, the robust controller determines its control action, trying to correct the vertical position of the bar's free end. The fuzzy controller also provides a control signal in an attempt to eliminate the tracking error; for this, it needs this error signal and its derivative. The control signal which actually will act on the plant will be the weighted sum of signals produced by the controllers. The degree of participation of each control action is determined by the supervisor, which uses as well as the fuzzy controller, the error information and its derivative. According to the control signal, the servo-actuator will provide vertical displacements to bar's center, to correct the tracking error.

The two output functions used are the same presented in [1]. They are described in Table 3.

Function name	Parameters [e(t) de(t)/d(t) 1]x[t s 1]^T
LTR	[0 0 0]
FUZ	[0 0 1]

Table 3. Output functions' parameters of the fuzzy supervisor

So, when supervisor output is null, only the robust controller will actuate on the plant, when supervisor output is equal to one, only the fuzzy controller will actuate, for intermediate outputs a combination of those controllers' signals will be applied on the plant.

The supervisor's input membership functions were tuned by a genetic algorithm using the square wave reference and the two controllers. Its evaluation function is given by:

$$f_{ev}(ind.) = t_{r1} + t_{r2} + t_{r3} + o_{s1} + |o_{s2}| + o_{s3} + t_{s1} + t_{s2} + t_{s3} + e_m^2$$
$$+0,01u_{max1} - 0,01u_{min1} + 0,01u_{max2} - 0,01u_{min2} + 0,01u_{max3} - 0,01u_{min3} \tag{40}$$

There was no need to give greater weight to the mean square error and to the rise times, as was done for the tuning of the fuzzy controller, because from some tunings, the settling time and the overshoot became very small. The reduction of all performance descriptors along the supervisor tuning was so high that it was necessary to assign lower weights to control signals peaks above the saturation of the servo actuator, to avoid favoring a performance criterion and neglect others.

Figures 23 and 24, shows the fuzzy supervisor optimized membership functions.

The rule base of the supervisor was not optimized by genetic algorithm. It was the same used in [1], as shown in Table 4.

As mentioned the results obtained with the optimized hierarchical fuzzy controller will be compared with the ones obtained by the non-optimized one (presented in [1]). On tracking a 0.1m amplitude step reference, the optimized hierarchical fuzzy controller has satisfied all performance criteria established and presented a more rapid response than the system controlled by the non-optimized hierarchical fuzzy controller. The rise time from 0 to 100% of the reference was approximately 0.22s, which is half the one obtained in [1]. The overshoot was 3.9% in [1] it was 7%. The settling time for (± 5%) was 0.22s, less than half that was obtained in [1]. The control signal generated by the optimized hierarchical fuzzy controller to track this reference had lower levels than the ones generated by the non-optimized hierarchical fuzzy controller. The optimized hierarchical fuzzy controller has used the fuzzy controller for less time, it is because the optimized fuzzy controller provide a faster response than the designed in [1]. Also the transition between controllers was softer with the optimized system. Figures 25, 26 and 27 shows the results obtained with those two structures on the reference tracking in disturbances absence.

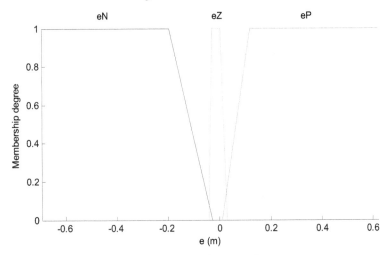

Figure 23. Fuzzy supervisor optimized membership functions of error input

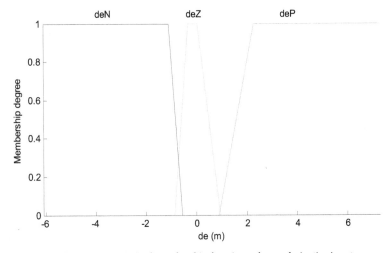

Figure 24. Fuzzy Supervisor optimized membership functions of error derivative input

		e (t)		
		eN	eZ	eP
de(t)/dt	deN	FUZ	FUZ	FUZ
	deZ	FUZ	LTR	FUZ
	deP	FUZ	FUZ	FUZ

Table 4. Rule base of supervisor

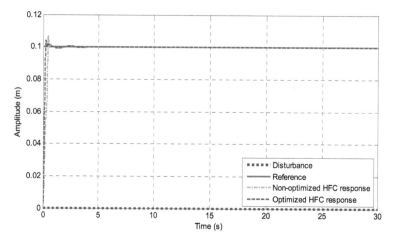

Figure 25. Comparison of the two hierarchical controllers in tracking a step reference

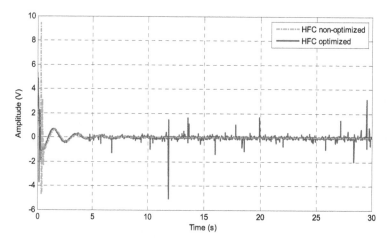

Figure 26. Comparison of control signals generated by the two hierarchical controllers in tracking a step reference

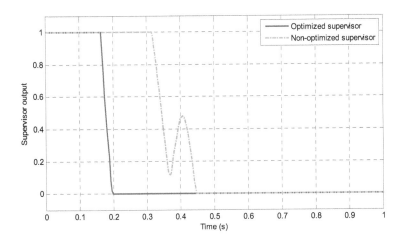

Figure 27. Comparison of signals generated by the two supervisors in tracking a step reference

The performance of the optimized HFC was tested on a step reference tracking, in the presence of white noise with 0.02m peak to peak. Figures 28 and 29 show, again, the best performance of the system controlled by the optimized HFC.

To finalize the comparisons, the system was tested on tracking a square wave reference. As expected, a better performance was obtained using the optimized HFC.

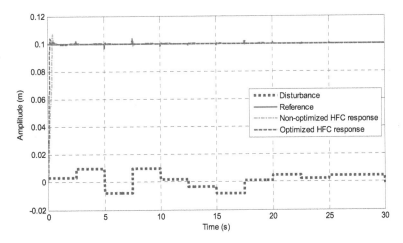

Figure 28. Comparison of the two HFC in tracking a step reference under disturbance

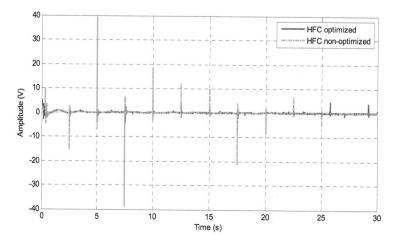

Figure 29. Comparison of control signals generated by the two HFC in tracking a step reference under disturbance

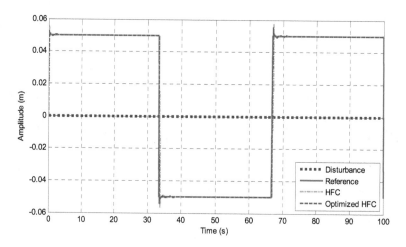

Figure 30. Comparison of the two HFC in tracking a square wave reference

As could be seen on Figure 31, both hierarchical controllers extrapolated the actuator saturation limits, as it was punctual the use of a saturator may not affect the system performance.

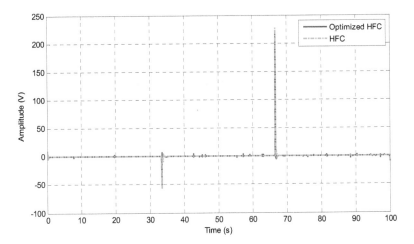

Figure 31. Comparison of control signals generated by the two HFC in tracking a square wave reference

Again the optimized supervisor has used less the fuzzy controller than the non-optimized supervisor.

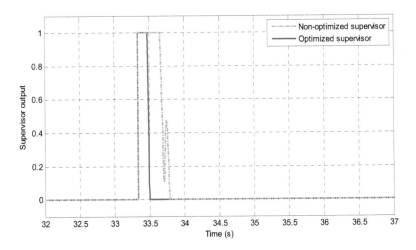

Figure 32. Comparison of signals generated by the two supervisors in tracking a square wave reference

6. Conclusion

One of the main advantages of hierarchical control is to combine different techniques. It allows the supervisor to take the best of each technique.

The results showed the advantages of using genetic algorithms, such as: making automatic tuning of fuzzy components of the HFC, greatly simplifying the design and allowing the obtaining of optimal controllers and supervisors, which is impossible via manual tuning.

As can be seen, the controllers were designed, relaxing some conflicting performance criteria: on the robust controller design the efforts were concentrated to obtain a rapid response and a rapid accommodation, in tracking references and in disturbance rejection, not worrying about the control signal amplitude, for references tracking. In fuzzy controller design the efforts were concentrated to obtain a rapid response and smaller control signals, but no major requirements for rapid accommodation, which had already been achieved by the robust controller; this way all performance requirements were satisfied through the use of the hierarchical fuzzy controller.

With the use of hierarchical control, the controller design becomes simpler because they are more specific, they do not have to meet conflicting performance criteria.

As a suggestion for future projects can be verified: other control techniques for vibration suppression and tracking reference; new ways to optimize the components of HFC; using more controllers in the composition of HFC; other methods for supervisor project; a better configuration of the proposed genetic algorithms.

Author details

Carlos André Guerra Fonseca
Informatics Department, Rio Grande do Norte State University, Natal, Brazil
Computing Engineering and Automation Department, Rio Grande do Norte Federal University, Natal, Brazil

Fábio Meneghetti Ugulino de Araújo
Computing Engineering and Automation Department, Rio Grande do Norte Federal University, Natal, Brazil

Marconi Câmara Rodrigues
Science and Technology School, Rio Grande do Norte Federal University, Natal, Brazil

7. References

[1] Araújo FMU (2002) Automatic Intelligent Controllers with Applications in Mechanical Vibration Isolation [PhD thesis]. Sao Jose dos Campos, Brazil: Aeronautical Technology Institute.

[2] Araújo FMU (1998) Electromechanical System for Vibration Active Control [Master thesis]. Joao Pessoa, Brazil: Paraiba Federal University.

[3] Araújo FMU and Yoneyama T (2001) Modeling and Control of an Electromechanical Device for Vibrations Active Control. Proceedings of the II National Seminar on Control and Automation. pp.15. Salvador, Brazil.

[4] Cruz JJ (1988) Contribution to the Study of Robust Stability for Nonlinear Multivariable Regulators [PhD thesis]. Sao Jose dos Campos, Brazil: National Institute of Space Research.

[5] Cruz JJ (1996) Multivariable Robust Control. Sao Paulo: Publisher from Sao Paulo University.

[6] Doyle JC and Stein G (1981) Multivariable Feedback Design: Concepts for a Classical/Modern Synthesis. IEEE Transactions on Automatic Control, Vol. AC-26, No. 1, pp. 4-16.

[7] Zadeh LA (1965) Fuzzy set. Fuzzy Sets, Information and Control, 8, pp.338-353.

[8] Lee CC (1990) Fuzzy Logic in Control Systems: Fuzzy Logic Controller (Part I). IEEE Transactions on Systems, Man and Cybernetics, 20(2): 404-418.

[9] Driankov D, Hellendoorn H, Reinfrank M (1993) An Introduction to Fuzzy Control. New York: Springer-Verlag. 316 p.

[10] Dutta S (1993) Fuzzy Logic Applications: Technological and Strategic Issues. IEEE Transactions on Engineering Management, 40(3): 237-254.

[11] Karr CL, Gentry EJ (1993) Fuzzy Control of Ph Using Genetic Algorithms. IEEE Transactions on Fuzzy Systems, 1(1): 46-53.

[12] Chiu S, Chand S (1993) Adaptive Traffic Signal Control Using Fuzzy Logic. Proceedings of the 2nd IEEE International Conference on Fuzzy Systems; March 1993; San Francisco, USA. Piscataway: Institute of Electrical and Electronics Engineers.

[13] Castro JL (1995) Fuzzy Logic Controllers are Universal Approximators. IEEE Transactions on Systems, Man and Cybernetics, 25(4): 629-635.

[14] Guerra R, Sandri S, Souza MLO (1997) Autonomous Control of Satellites Altitude Using Fuzzy Logic. In: III Brazilian Symposium on Intelligent Automation, pp.337-342, Vitoria, Brazil.

[15] Kosko B (1992) Neural Networks and Fuzzy System. Englewood Cliffs: Prentice Hall. 449 p.

[16] Wang L, Mendel J M (1992) Generating Fuzzy Rules by Learning from Examples. IEEE Transactions on Systems, Man, and Cybernetics, 22(6): 1414-1427.

[17] Jang JSR (1993) ANFIS: Adaptive-Network-Based Fuzzy Inference System. IEEE Transactions on Systems, Man, and Cybernetics, 23(3): 665-685.

[18] Lin CT (1995) A Neural Fuzzy Control System with Structure and Parameter Learning. Fuzzy Sets and Systems, (70): 183-212.

[19] Sandri AS, Correa C (1999) Fuzzy Logic. In: V Neural Networks School, São José dos Campos, Brazil. pp.c073-c090.

[20] Shaw IS, Simões MG (1999) Fuzzy Control and Modeling. São Paulo: FAPESP, Editora Edgard Blücher LTDA.

[21] Tsoukalas LH, Uhrig RE (1997) Fuzzy and Neural Approaches in Engineering. New York: Publication, John Wiley & Sons.

[22] Fonseca CAG, Araújo FMU, Maitelli AL, Medeiros AV (2003) Genetic Algorithms for Optimization of a Fuzzy Controller for Vibration Suppression. In: VI Brazilian Symposium on Intelligent Automation, pp. 959-963. Bauru, Brazil.

[23] Goldbarg MC and Goldbarg EFG (2005) Evolutionary Computation. In: VIII Neural Networks School. Natal, Brazil.

[24] Holland JH (1970) Robust Algorithms for Adaptation Set in a General Formal Framework. In: IEEE Symposium on Adaptive Processes Decision and Control, 17. Proceedings of the XVII IEEE Symposium on Adaptive Processes Decision and Control.

[25] Holland JH (1975) Adaptation in Natural and Artificial Systems. Ann Arbor: Michigan University Press.

Chapter 3

Fuzzy Control Systems: LMI-Based Design

Morteza Seidi, Marzieh Hajiaghamemar and Bruce Segee

Additional information is available at the end of the chapter

1. Introduction

This chapter describes widespread methods of model-based fuzzy control systems. The subject of this chapter is a systematic framework for the stability and design of nonlinear fuzzy control systems. We are trying to build a bridge between conventional fuzzy control and classic control theory. By building this bridge, the strong well developed tools of classic control could be used in model-based fuzzy control systems

Model-based fuzzy control, with the possibility of guaranteeing the closed loop stability, is an attractive method for control of nonlinear systems. In recent years, many studies have been devoted to the stability analysis of continuous time or discrete time model based fuzzy control systems (Takagi & Sugeno, 1985; Rhee & Won, 2006; Chen et al., 1993; Wang et al., 1996; Zhao et al., 1996; Tanaka & Wang, 2001; Tanaka et al., 2001). Among such methods, the method of Takagi-Sugeno (Takagi & Sugeno, 1985) has found many applications for modelling complex nonlinear systems (Tanaka & Sano, 1994;Tanaka & Kosaki, 1997;Li et al., 1998). The concept of sector nonlinearity (Kawamoto et al., 1992) provided means for exact approximation of nonlinear systems by fuzzy blending of a few locally linearized subsystems. One important advantage of using such a method for control design is that the closed-loop stability analysis, using the Lyapunov method, becomes easier to apply. Various stability conditions have been proposed for such systems (Tanaka &Wang, 2001), (Ting, 2006), where the existence of a common solution to a set of Lyapunov equations is shown to be sufficient for guaranteeing the closed-loop stability. Some relaxed conditions are also proposed in (Kim & Lee, 2000; Ding et al, 2006; Fang et al., 2006, Tanaka & Ikeda, 1998). Parallel Distributed Compensator (PDC) is a generalization of the state feedback controller to the case of nonlinear systems, using the Takagi-Sugeno fuzzy model (Wang et al., 1996). This method is based on partitioning nonlinear system dynamics into a number of linear subsystems, for which state feedback gains are designed and blended in a fuzzy sense. Takagi-Sugeno model and parallel distributed compensation have been used in many applications successfully (Sugeno & Kang, 1986, Lee et al., 2006, Hong & Langari, 2000, Bonissone et al.,

1995). The Linear Matrix Inequality (LMI) technique offers a numerically tractable way to design a PDC controller with objectives such as stability (Wang et al.,1996; Ding et al, 2006; Fang et al., 2006; Tanaka & Sugeno 1992), H∞ control (Lee et al., 2001), H2 control (Lin & Lo, 2003), pole-placement (Jon et al, 1997; Kang & Lee, 1998), and others (Tanaka & Wang, 2001).

2. Takagi-Sugeno fuzzy model

The main idea of the Takagi-Sugeno fuzzy modeling method is to partition the nonlinear system dynamics into several locally linearized subsystems, so that the overall nonlinear behavior of the system can be captured by fuzzy blending of such subsystems. The fuzzy rule associated with the i-th linear subsystem for the continuous fuzzy system and the discrete fuzzy system, can then be defined as

Continuous fuzzy system

$$\text{Rule i :IF } Z_1(t) \text{ is } M_{i1} \dots \text{ and } Z_1(t) \text{ is } M_{il}$$

$$\text{THEN} \quad \begin{cases} \dot{x}(t) = A_i x(t) + B_i u(t) \\ \quad y(t) = C_i x(t) \end{cases} \quad i=1,2,\dots,r \tag{1}$$

Discrete Fuzzy System

$$\text{Rule i :IF } Z_1(t) \text{ is } M_{i1} \dots \text{ and } Z_1(t) \text{ is } M_{il}$$

$$\text{THEN} \quad \begin{cases} x(t+1) = A_i x(t) + B_i u(t) \\ \quad y(t) = C_i x(t) \end{cases} \quad i=1,2,\dots,r \tag{2}$$

where, $x(t) \in R^n$ is the state vector, $u(t) \in R^m$ is the input vector, $A_i \in R^{n \times n}$, $B_i \in R^{n \times m}$, $C_i \in R^{q \times n}$; $\{z_1(t), z_2(t),\dots,z_p(t)\}$ are nonlinear functions of the state variables obtained from the original nonlinear equation, and $M_{ij}(z_i)$ are the degree of membership of $z_i(t)$ in a fuzzy set M_{ij}. Whenever there is no ambiguity, the time argument in $z(t)$ is dropped. The overall output, using the fuzzy blend of the linear subsystems, will then be as follows:

Continuous fuzzy system

$$\dot{X} = \frac{\sum_{I=1}^{R} w_1(z)\{A_i x(t) + B_i u(t)\}}{\sum_{i=1}^{r} w_i(z)} = \sum_{i=1}^{r} h_1(z)\left(A_i x(t) + B_i u(t)\right)$$

$$y(t) = \frac{\sum_{i=1}^{r} w_1(z) C_i x(t)}{\sum_{i=1}^{r} w_1(z)} = \sum_{i=1}^{r} h_1(z) C_i x(t) \tag{3}$$

Discrete Fuzzy System

$$x(t+1) = \frac{\sum_{i=1}^{r} \omega_i\left(z(t)\right)\left\{A_i x(t) + B_i u(t)\right\}}{\sum_{i=1}^{r} \omega_i\left(z(t)\right)}$$

$$= \sum_{i=1}^{r} h_i\left(z(t)\right)\left(A_i x(t) + B_i u(t)\right)$$

$$y(t) = \frac{\sum_{i=1}^{r} \omega_i\left(z(t)\right) C_i x(t)}{\sum_{i=1}^{r} \omega_i\left(z(t)\right)}$$ (4)

$$= \sum_{i=1}^{r} h_i\left(z(t)\right) C_i x(t)$$

Where

$$w_1(z) = \prod_{j=1}^{i} M_{ij}\left(z_j\right)$$

$$h_1(z) = \frac{w_1(z)}{\sum_{i=1}^{r} w_1(z)}$$ (5)

It is also true, for all t, that

$$\begin{cases} \sum_{i=1}^{r} w_1(z) > 0, \\ w_1(z) \ge 0, i = 1, 2, \ldots\ldots, r \end{cases}$$

2.1. Building a fuzzy model

There are generally three approaches to build the fuzzy model: "sector nonlinearity," "local approximation," or a combination of the two.

2.1.1. Sector nonlinearity

Figure 1 illustrates the concept of global and local sector nonlinearity. Suppose the original nonlinear system satisfies the sector non-linearity condition (Kawamoto et al., 1992, as cited in Tanaka & Wang, 2001), i.e., the values of nonlinear terms in the state-space equation remain within a sector of hyper-planes passing through the origin. This model guarantees the stability of the original nonlinear system under the control law. A function Φ: R→R is said to be sector [a,c] if for all x∈R, y= Φ(x) lies between $b_1 x$ and $b_2 x$.

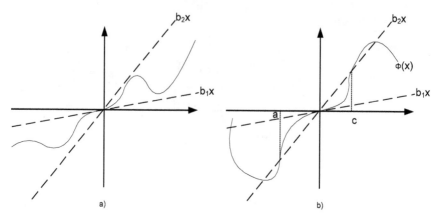

Figure 1. a) Global sector nonlinearity, b) Local sector nonlinearity

Example 1

The well-known nonlinear control benchmark, the ball-and-beam system is commonly used as an illustrative application of various control methods (Wang & Mendel, 1992) depicted in figure 2. Let $x_1(t)$ and $x_2(t)$ denote the position and the velocity of the ball and let $x_3(t)$ and $x_4(t)$ denote the angular position and the angular velocity of the beam Then, the system dynamics can be described by the following state-space equation

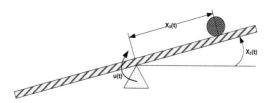

Figure 2. The ball and beam system

$$\dot{x}(t) = f(x(t)) + g(x(t))u(t)$$

Where   (6)

$$f(x) = \begin{bmatrix} x_2(t) \\ B(x_1(t)x_4^2(t) - G\sin(x_3(t))) \\ x_4(t) \\ 0 \end{bmatrix} \quad \text{and} \quad g(x) = \begin{bmatrix} 0 \\ 0 \\ 0 \\ 1 \end{bmatrix}$$

Where $x(t) = \begin{bmatrix} x_1(t) & x_2(t) & x_3(t) & x_4(t) \end{bmatrix}^T$ and u(t) is torque.

$\sin(x_3)$ and $x_1x_4^2$ are nonlinear terms in the state-space equation. We define $z_1 = \sin(x_3)$ and $z_2 = x_1x_4^2$. Assume $x_3 \in \left[-\frac{\pi}{2}, \frac{\pi}{2}\right]$ and $x_1x_4 \in \left[-d \ d\right]$ as the region within which the system will operate. Figure 3 shows that $z_1(t) = \sin(x_3(t))$ and its local sector operating region. The sector $[b_1, b_2]$ consists of two lines b_1x_1 and b_2x_1, where the slopes are $b_1 = 1$ and $b_2 = \frac{2}{\pi}$. It follows that

$$\left|\frac{2}{\pi}x\right| \le |sin(x)| \le |x|,$$

$$-dx_4 \le x_1x_4^2 \le dx_4.$$

(7)

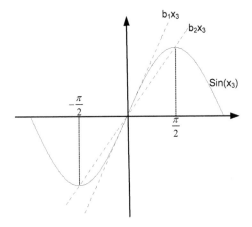

Figure 3. $sin\big(x_3(t)\big)$ and its local sector

We present $sin\big(x_3(t)\big)$ is represented as follows:

$$z_1 = sin\big(x_3(t)\big) = \left(\sum_{i=1}^{2} M_i\big(z_1(t)\big)b_i\right)x_3(t)$$

(8)

From the property of membership functions $\Big[M_1\big(z_1(t)\big) + M_2\big(z_1(t)\big) = 1\Big]$, we can obtain the membership functions

$$M_1\big(z_1(t)\big) = \begin{cases} \dfrac{z_1(t) - \big(\frac{2}{\pi}\big)S\sin(z_1(t))}{\big(1 - \frac{2}{\pi}\big)\sin^{-1}(z_1(t))} & z_1(t) \ne 0 \\ 1, & otherwise. \end{cases}$$

$$M_2\big(z_1(t)\big) = \begin{cases} \dfrac{\sin^{-1}(z_1(t)) - z_1(t)}{\big(1 - \frac{2}{\pi}\big)\sin^{-1}(z_1(t))} & z_1(t) \ne 0 \\ 0, & otherwise. \end{cases}$$

(9)

Similarly we obtain membership functions associated with $z_2(t) = x_1(t)x_4(t)$. Assume $max(z_2(t)) = d = \alpha_1$ and $min(z_2(t)) = -d = \alpha_2$ we have:

$$z_2(t) = x_1(t)x_4(t) = \left(\sum_{i=1}^{2} N_i\left(z_2(t)\right) b_i \right) \alpha_i \tag{10}$$

$$N_1\left(z_2(t)\right) = \frac{z_2(t) - \alpha_2}{\alpha_1 - \alpha_2},$$

$$N_2\left(z_2(t)\right) = \frac{\alpha_1 - z_2(t)}{\alpha_1 - \alpha_2}, \tag{11}$$

The exact TS-fuzzy model-based dynamic system of the ball and beam system can be obtained as following:

$$\begin{bmatrix} \dot{x}_1(t) \\ \dot{x}_2(t) \\ \dot{x}_3(t) \\ \dot{x}_4(t) \end{bmatrix} = \sum_{i=1}^{2}\sum_{j=1}^{2} M_i(z_1(t)) N_j(z_2(t)) \times \left(\begin{bmatrix} 0 & 1 & 0 & 0 \\ 0 & 0 & -Gb_i & D\alpha_j \\ 0 & 0 & 0 & 1 \\ 0 & 0 & 0 & 0 \end{bmatrix} \begin{bmatrix} x_1(t) \\ x_2(t) \\ x_3(t) \\ x_4(t) \end{bmatrix} + \begin{bmatrix} 0 \\ 0 \\ 0 \\ 1 \end{bmatrix} u(t) \right) \tag{12}$$

The fuzzy model has the following 4 rules:

Rule 1: if $z_1(t)$ is M_1 and $z_2(t)$ is N_1
   Then $\dot{x}(t) = A_1 x(t) + B_1 u(t)$,
Rule 2: if $z_1(t)$ is M_1 and $z_2(t)$ is N_2
   Then $\dot{x}(t) = A_2 x(t) + B_2 u(t)$,
Rule 3: if $z_1(t)$ is M_2 and $z_2(t)$ is N_1
   Then $\dot{x}(t) = A_3 x(t) + B_3 u(t)$,
Rule 4: if $z_1(t)$ is M_2 and $z_2(t)$ is N_2
   Then $\dot{x}(t) = A_4 x(t) + B_4 u(t)$

$$\tag{13}$$

Where

$$A_1 = \begin{bmatrix} 0 & 1 & 0 & 0 \\ 0 & 0 & -Gb_1 & D\alpha_1 \\ 0 & 0 & 0 & 1 \\ 0 & 0 & 0 & 0 \end{bmatrix}, A_1 = \begin{bmatrix} 0 & 1 & 0 & 0 \\ 0 & 0 & -Gb_1 & D\alpha_2 \\ 0 & 0 & 0 & 1 \\ 0 & 0 & 0 & 0 \end{bmatrix},$$

$$A_1 = \begin{bmatrix} 0 & 1 & 0 & 0 \\ 0 & 0 & -Gb_2 & D\alpha_1 \\ 0 & 0 & 0 & 1 \\ 0 & 0 & 0 & 0 \end{bmatrix}, A_1 = \begin{bmatrix} 0 & 1 & 0 & 0 \\ 0 & 0 & -Gb_2 & D\alpha_2 \\ 0 & 0 & 0 & 1 \\ 0 & 0 & 0 & 0 \end{bmatrix}$$

$$B_1 = B_2 = B_3 = B_4 = B = \begin{bmatrix} 0 \\ 0 \\ 0 \\ 1 \end{bmatrix}, z_1 = \sin(x_3) \text{ and } z_2 = x_1 x_4$$

2.1.2. Local approximation

The original system can be partitioned into subsystems by approximation of nonlinear terms about equilibrium points. This approach can have fewer rules and of course less complexity but it cannot guarantee the stability of the original system under the controller. Usually in this approach, construction of a fuzzy membership function requires knowledge of the behavior of the original system and of course different types of membership functions can be selected.

3. Parallel distributed compensation

Parallel distributed compensation (PDC) is a model-based design procedure introduced in (Wang et al,. 1995). Using the Takagi-Sugeno fuzzy model, a fuzzy combination of the stabilizing state feedback gains, $F_i, i = 1, 2, ..., r$, associated with every linear subsystem is used as the overall state feedback controller. The general structure of the controller is then as

If $z_1(t)$ is M_{i1}, and $z_2(t)$ is M_{i2}, ...m, and $z_p(t)$ is M_{ip} then $u = -F_i x(t), i = 1, 2, ..., r$ (14)

The output of the controller is represented by

$$u = -\frac{\sum_{i=1}^{r} \omega_i(z) F_i x(t)}{\sum_{i=1}^{r} \omega_i} = -\sum_{i=1}^{r} h_i(z) F_i x(t).$$ (15)

The Takagi-Sugeno model and the Parallel Distributed Compensation have the same number of fuzzy rules and use the same membership functions.

4. Stability conditions and control design

4.1. LMI

A variety of problems arising in system and control theory can be reduced to a few standard convex or quasi-convex optimization problems involving linear matrix inequalities (LMIs). Lyapunov published his theory in 1890 and showed that $\frac{d}{dt} x(t) = Ax(t)$ is stable if and only if there exists a positive-definite matrix P such that $A^T P + PA < 0$. The Lypanov inequality, $P > 0$ and $A^T P + PA < 0$ is a form of an LMI.

An LMI has the form

$$F(x) \triangleq F_0 + \sum_{i=1}^{m} x_i F_i > 0,$$ (16)

Where $F_i \in R^{n \times n}, i = 0,...,m$ are the given symmetric matrices and $x \in R^m$ is the variable and the inequality symbol shows that $F(x)$ is positive definite (Boyd, 1994).

4.2. Stability conditions

There are a large number of works on stability conditions and control design of fuzzy systems in the literature. A sufficient stability condition for ensuring stability of PDC was derived by Tanaka and Sugeno (Tanaka & Sugeno, 1990; 1992).

By substituting the controller output (15) into the TS model for the continuous fuzzy control (4), we have:

$$\dot{x}(t) = \sum_{i=1}^{r} \sum_{j=1}^{r} h_i(z(t)) h_j(z(t)) \{ A_i - B_i F_j \} x(t) \tag{17}$$

or

$$\dot{x}(t) = \sum_{j=1}^{r} h_i(z(t)) h_i(z(t)) G_{ii} x(t)$$
$$+ 2 \sum_{i=1}^{r} \sum_{i<j} h_i(z(t)) h_j(z(t)) \left\{ \frac{G_{ij} + G_{ji}}{2} \right\} x(t) \tag{18}$$

where $G_{ij} = A_i - B_i F_j$, Similarly for the discrete fuzzy system we have

$$x(t+1) = \sum_{i=1}^{r} \sum_{j=1}^{r} h_i(z(t)) h_j(z(t)) \{ A_i - B_i F_j \} x(t) \tag{19}$$

or

$$\dot{x}(t+1) = \sum_{j=1}^{r} h_i(z(t)) h_i(z(t)) G_{ii} x(t)$$
$$+ 2 \sum_{i=1}^{r} \sum_{i<j} h_i(z(t)) h_j(z(t)) \left\{ \frac{G_{ij} + G_{ji}}{2} \right\} x(t) \tag{20}$$

Theorem 1: The equilibrium of the continuous fuzzy system (3) with u(t) = 0 is globally asymptotically stable if there exists a common positive definite matrix P such that

$$A_i^T P + P A_i < 0, \quad i = 1,2,...,r \tag{21}$$

that is, a common P has to exist for all subsystems.

Theorem 2: The equilibrium of the discrete fuzzy system (4) with u(t) = 0 is globally asymptotically stable i f there exists a common positive definite matrix P such that

$$A_i^T P A_i - P < 0, \quad i = 1,2,...,r \tag{22}$$

that is, a common P has to exist for all subsystems.

The stability of the closed loop system can be derived by using theorem 1 and 2.

Theorem 3: The equilibrium of the continuous fuzzy control system described by (18) is globally asymptotically stable if there exists a common positive definite matrix P such that

$$G_{ii}^T P + PG_{ii} < 0,$$

$$\left(\frac{G_{ij}+G_{ji}}{2}\right)^T P + P\left(\frac{G_{ij}+G_{ji}}{2}\right) \leq 0, \tag{23}$$

$$i < j \text{ s.t. } h_i \cap h_j \neq \phi \tag{24}$$

Theorem 4: The equilibrium of the discrete fuzzy control system described by (20) is globally asymptotically stable if there exists a common positive definite matrix P such that

$$G_{ii}^T PG_{ii} - P < 0,$$

$$\left(\frac{G_{ij}+G_{ji}}{2}\right)^T P\left(\frac{G_{ij}+G_{ji}}{2}\right) - P \leq 0, \tag{25}$$

$$i < j \text{ s.t. } h_i \cap h_j \neq \phi \tag{26}$$

4.3. Stable controller design

By using the following conditions, the solution of the LMI problem for continuous and discrete fuzzy systems gives us the state feedback gains F$_i$ and the matrix P (if the problem is solvable).

Consider a new variable $X = P^{-1}$ then the stable fuzzy controller design problem is:

Continuous fuzzy system

Find $X > 0$ and M_i, $i = 1,2,...,r$

$$-XA_i^T - A_i X + M_i^T B_i^T + B_i M_i > 0,$$
$$-XA_i^T - A_i X - XA_j^T - A_j X$$
$$+ M_j^T B_i^T + B_i M_j + M_i^T B_j^T + B_j M_i \geq 0. \tag{27}$$

$$X = P^{-1} \quad i < j \text{ s.t. } h_i \cap h_j \neq \phi \tag{28}$$

The conditions (27) and (28) gives us a positive definite matrix X and M_i (or that there is no solution). From the solution X and M_i, a common P and the feedback gains can be found as:

$$P = X^{-1}, \; F_i = M_i X^{-1} \tag{29}$$

Similarly for a discrete fuzzy system the design problem is

Find $X > 0$ and M_i, $i = 1,2,...,r$

$$X - \left(A_i X - B_i M_i \right)^T X^{-1} \left(A_i X - B_i M_i \right) > 0,$$

$$X - \frac{1}{4} X \left(A_i X - B_i M_i + A_j X - B_j M_i \right)^T X^{-1} \tag{30}$$

$$\times \left(A_i X - B_i M_j + A_j X - B_j M_i \right) X \geq 0.$$

4.4. Decay rate

Decay rate is associated with the speed of response. The decay rate fuzzy controller design helps to find feedback gains that provide better setteling time (Tanaka et al,. 1996; 1998a; 1998b).

Continuous fuzzy system: The condition that $\dot{V}\left(x(t)\right) \leq -2\alpha V\left(x(t)\right)$ (Ichikawa et al, 1993, as cited in Tanaka & Wang, 2001) for all $x(t)$ can be written as

$$G_{ii}^T P + P G_{ii} + 2\alpha P < 0$$

$$\left(\frac{G_{ij} + G_{ji}}{2} \right)^T P + P \left(\frac{G_{ij} + G_{ji}}{2} \right) + 2\alpha P \leq 0 \tag{31}$$

Where

$$G_{ij} = A_i - B_i F_i, \; \alpha > 0 \text{ and } i < j \text{ s.t. } h_i \cap h_j \neq \phi \tag{32}$$

Therefore, by solving the following generalized eigenvalue minimization problem in X, the largest lower bound on the decay rate that can be found by using a quadratic Lyapunov function:

maximize α subject to

$$X > 0,$$

$$-X A_i^T + A_i X + M_i^T B_i^T + B_i M_i - 2\alpha X > 0,$$

$$-X A_i^T - A_i X - X A_j^T - A_j X + M_j^T B_j^T + B_i M_j \tag{33}$$

$$+ M_i^T B_j^T + B_j M_i - 4\alpha X > 0,$$

$$i < j \text{ s.t. } h_i \cap h_j \neq \varphi, \text{ where } X = P^{-1}, \quad M_i = F_i X. \tag{34}$$

Similarly for a discrete fuzzy system:

The condition that $\Delta V\big(x(t)\big) \le \big(\alpha^2 - 1\big)V\big(x(t)\big)$ (Ichikawa et al, 1993, as cited in Tanaka & Wang, 2001) for all $x(t)$ can be written as

$$G_{ii}^T P G_{ii} - \alpha^2 P < 0,$$

$$\left(\frac{G_{ij} + G_{ji}}{2}\right)^T P \left(\frac{G_{ij} + G_{ji}}{2}\right) - \alpha^2 P \le 0 \tag{35}$$

$$i < j \text{ s.t. } h_i \cap h_j \ne \phi \text{ and } \alpha < 1 \tag{36}$$

The generalized eigenvalue minimization can be found in (Tanaka & Wang, 2001).

4.5. Constraint on control

Theorem 5: Assume that the initial condition x(0) is known. The constraint $\|u(t)\|_2 \le \mu$ is satisfied at all times $t \ge 0$ if the LMIs

$$\begin{bmatrix} 1 & x(0)^T \\ x(0) & X \end{bmatrix} \ge 0$$

$$\begin{bmatrix} X & M_i^T \\ M_i & \mu^2 I \end{bmatrix} \ge 0 \tag{37}$$

Hold, where $X = P^{-1}$ and $M_i = F_i X$.

The above LMI design conditions depend on the initial states. Thus, if the initial states $x(0)$ change, this means that the feedback gains Fi must be again determined. To overcome this disadvantage, modified LMI constraints on the control input have been developed, where $x(0)$ is unknown but the upper bound ϕ of $\|x(t)\|$ is known, i.e., $\|x(t)\| \le \phi$.

Theorem 6: Assume that $\|x(t)\| \le \phi$, where x(0) is unknown but the upper bound φ is known. Then,

$$x^T(0)X^{-1}x(0) \le 1 \text{ if } \phi^2 I \le X, \tag{38}$$

Where $X = P^{-1}$

Proofs of theorem 1 and 2 are given in (Tanaka & Wang, 2001)

4.6. Performance-oriented parallel distributed compensation

In the modified PDC proposed in (Seidi & Markazi, 2011), unlike the conventional PDC, state feedback gains associated with every linear subsystem, are not assumed fixed. Instead, based on some pre-specified performance criteria, several feedback gains are designed and used for every subsystem. The overall gain associated with each of the subsystems, is then determined by a fuzzy blending of such gains, so that a better closed-loop performance can be achieved. The required membership functions are chosen based on some pre-specified performance indices, for example, a faster response or a smaller control input. In general, the rest of the method for calculating the overall state feedback gain remains similar to the conventional PDC method, as in (14) and (15). Figure 4, depicts the general framework for the proposed method, through which and depending on various performance criteria, different characteristics for the controller can be specified. For example, two different feedback gains could be designed for a typical subsystem; one providing a lower control input with a longer settling time response, and the other a faster response but with a larger control input. The idea is then to select the overall feedback gain for this subsystem as a weighted sum of such gains, where the weights are appropriately adjusted, in a fuzzy sense, during the time evolution of the system response, so that as a whole, a faster response with a lower control input can be achieved. For this purpose, when the magnitude of the control input becomes large, the relative weight of the first feedback gain is increased, so that the magnitude of the control input is kept within the permissible limits. On the other hand, when the control input is well below the permissible limit, the weight of the second feedback gain is increased, for a faster response. The dynamics of the resulting closed-loop control system can be analyzed as follows:

Consider the following Takagi–Sugeno model of the plant

$$\dot{x} = \sum_{i=1}^{r} h_1(z)\{A_i x(t) + B_i u(t)\} \tag{39}$$

The following structure is proposed for the fuzzy controller rules

i th rule : If $Z_1(t)$ is M_{i1} and $Z_2(t)$ is M_{i2},........,$Z_p(t)$ is M_{ip}, $J(t)$ is H_{i1},....and $J(t)$ is H_{iq}

$$\text{then } u_i(t) = \left\{\sum_{n=1}^{q} m_{in}(J(t))K_{in}\right\} x(t) \tag{40}$$

Where $i = 1,2,...,r$, q_i is the number of gain coefficients in the ith subsystem, m_{in} is the relevant membership degree for J(t), K_{in} is the nth state feedback gain associated with the ith subsystem, H_{iq} is the n th membership function for J(t), defined in the ith rule. Here $J(t)$ is a term depicting a selected performance index, for instance, if one wants to limit the magnitude of the control signal $u(t)$, then $J(t) \equiv |u(t)|$. Where the control input generated by the PDC controller is in the form of

$$u(t) = \sum_{i=1}^{r} h_i(z) u_i(t) = -\left\{\sum_{n=1}^{r} h_i(z) K_i\right\} x(t)$$

$$K_i = \sum_{n=1}^{q} m_{in}(J(t)) K_{in} \tag{41}$$

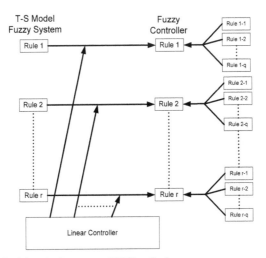

Figure 4. General methodology in the proposed PDC method

Lemma: The fuzzy control system (39), with the control strategy (41) is globally, asymptotically stable, if there exists a common positive definite matrix P such that

$$G_{iin}^{T}P + PG_{iin} < 0$$

$$\left(\frac{G_{ijn}+G_{jin}}{2}\right)^{T} P + P\left(\frac{G_{ijn}+G_{jin}}{2}\right) \leq 0 \qquad (42)$$

where $i < j$, $h_i \cap h_j \neq \phi$, $G_{ijn} = A_i - B_i K_{jn}$.

Example 2

Consider a single link robot with flexible joint as in Figure 5. This benchmark problem is introduced in (Spong et al., 1987).

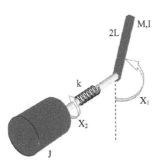

Figure 5. A single link robot with a flexible joint

The state space equations for the system of Figure 4 are

$$
\begin{cases}
\dot{x}_1 = x_3(t) \\
\dot{x}_2 = x_4(t) \\
\dot{x}_3 = \dfrac{1}{I}\left(k(x_2(t) - x_1(t)) - mgLsin(x_1(t))\right) \\
\dot{x}_4 = \dfrac{1}{J}\left(u(t) - k(x_2(t) - x_1(t))\right)
\end{cases}
\tag{43}
$$

In order to apply the PDC methodology, the fuzzy Takagi-Sugeno Model is developed first (Seidi & Markazi, 2008). The nonlinear expression $Z = sin(x_1(t))$, for $x_1(t) \in [-pi, pi]$, can be expressed as

$$
z = sin(x_1(t)) = \left(\sum_{i=1}^{2} M_i(z) b_i\right) x_1(t)
\tag{44}
$$

Where, $b_1 = 1, b_2 = 0$ and, hence, the membership functions for z are obtained as

$$
M_1(z) = \begin{cases} \dfrac{z}{Sin^{-1}z}, & z(t) \neq 0 \\ 1, & \text{Otherwise} \end{cases}
$$

$$
M_2(z) = \begin{cases} \dfrac{Sin^{-1}z - z}{Sin^{-1}z}, & z(t) \neq 0 \\ 1, & \text{Otherwise} \end{cases}
\tag{45}
$$

The resulting fuzzy model would then have the following fuzzy rules:

$$
\begin{aligned}
&Rule\ 1: \text{If } z(t) \text{ is } M_1(z), \text{then } \dot{x}(t) = A_1 x(t) + B_1 u(t) \\
&Rule\ 1: \text{If } z(t) \text{ is } M_2(z), \text{then } \dot{x}(t) = A_2 x(t) + B_2 u(t)
\end{aligned}
\tag{45}
$$

Where,

$$
A_1 = \begin{bmatrix} 0 & 0 & 1 & 0 \\ 0 & 0 & 0 & 1 \\ \dfrac{-k-mgLb_1}{I} & \dfrac{k}{I} & 0 & 0 \\ \dfrac{k}{J} & -\dfrac{k}{J} & 0 & 0 \end{bmatrix},\
A_2 = \begin{bmatrix} 0 & 0 & 1 & 0 \\ 0 & 0 & 0 & 1 \\ \dfrac{-k-mgLb_2}{I} & \dfrac{k}{I} & 0 & 0 \\ \dfrac{k}{J} & -\dfrac{k}{J} & 0 & 0 \end{bmatrix},
\tag{46}
$$

and

$$
B_1 = B_2 = B = \begin{bmatrix} 0,0,0,1 \end{bmatrix}^T.
\tag{47}
$$

Assume $k = 100 \ Nm / rad$, $g = 9.8 \ m / s^2$ and other parameters are assumed unity then we have

$$A_1 = \begin{bmatrix} 0 & 0 & 1 & 0 \\ 0 & 0 & 0 & 1 \\ -109.8 & 100 & 0 & 0 \\ 100 & -100 & 0 & 0 \end{bmatrix}, \ A_2 = \begin{bmatrix} 0 & 0 & 1 & 0 \\ 0 & 0 & 0 & 1 \\ 0 & 100 & 0 & 0 \\ 100 & -100 & 0 & 0 \end{bmatrix}, \ B = \begin{bmatrix} 0 \\ 0 \\ 0 \\ 1 \end{bmatrix},$$

Control Rule 1:
If $z(t)$ is $M_1(z)$, then $u(t) = -F_1 x(t)$

Control Rule 2:
If $z(t)$ is $M_2(z)$, then $u(t) = -F_2 x(t)$

(48)

The final output of the controller is

$$u(t) = -\sum_{i=1}^{2} h_i F_i x(t) = h_1 F_1 x(t) + h_2 F_2 x(t)$$

(49)

Case 1: Stable controller design

Using conditions (27) and (28) the stable controller can be obtained by solving below conditions

$$X > 0$$
$$\left[-X A_1 - A_1 X + M_1{}^T B^T + B M_1 \right] > 0,$$
$$\left[-X A_2 - A_2 X + M_2{}^T B^T + B M_2 \right] > 0,$$
$$\left[-X A_1{}^T - A_1 X - X A_2{}^T - A_2 X + M_2{}^T B^T + B M_2 + M_1{}^T B^T + B M_1 \right] > 0$$

(50)

Using the MATLAB LMI Control Toolbox we obtain

$$F_1 = [-495.76 \quad 668.96 \quad 14.112 \quad 47.388]$$
$$F_2 = [-497.23 \quad 671.34 \quad 14.356 \quad 47.552]$$

(51)

$$P = \begin{bmatrix} 42.1464 & -50.7108 & -1.5337 & -3.2007 \\ -50.7108 & 68.9721 & 2.4898 & 4.3456 \\ -1.5337 & 2.4898 & 0.2554 & 0.1719 \\ -3.2007 & 4.3456 & 0.1719 & 0.3527 \end{bmatrix}$$

Figures 6 and 7 show the response of the system and control effort, respectively.

Case 2: The decay rate

Using conditions (31) and (32) the stable controller can be obtained by solving the conditions:

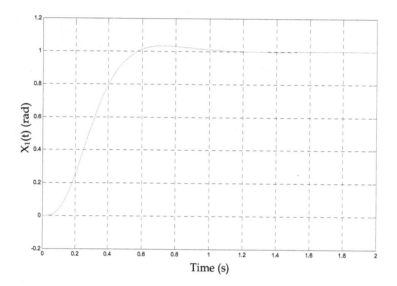

Figure 6. Response of flexible joint robots $x_1(t)$, case 1.

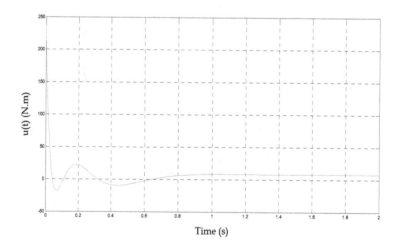

Figure 7. Control input for flexible joint robots, case 1.

$$[-X A_1^T - A_1 X + M_1^T B^T + B M_1 - 2\alpha X] > 0$$
$$[-X A_2^T - A_2 X + M_2^T B^T + B M_2 - 2\alpha X] > 0$$
$$[-X A_1^T - A_1 X - X A_2^T - A_2 X + M_2^T B^T$$
$$+B M_2 + M_1^T B^T + B M_1 - 4\alpha X] > 0$$

(52)

Considering $\alpha = 10$ and by using the MATLAB LMI Control Toolbox we obtain:

$$F_1 = [4108.8 \quad 6545.2 \quad 1271.3 \quad 127.77]$$
$$F_2 = [4066.9 \quad 6502.6 \quad 1261.7 \quad 127.1]$$

(53)

$$P = \begin{bmatrix} 36.5087 & 24.0140 & 6.2135 & 0.3352 \\ 24.0140 & 30.1341 & 6.3223 & 0.5013 \\ 6.2135 & 6.3223 & 1.4260 & 0.0995 \\ 0.3352 & 0.5013 & 0.0995 & 0.0099 \end{bmatrix}$$

Figures 8 and 9 show the response of the system and control effort, respectively.

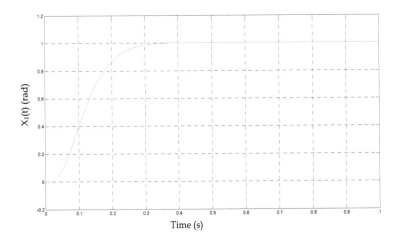

Figure 8. Response of flexible joint robots $x_1(t)$, case 2.

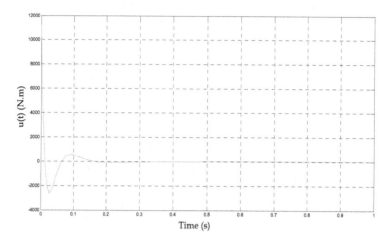

Figure 9. Control input for flexible joint robots, case 2.

Case 3: The decay rate with the constraint on the input

We design a stable fuzzy controller by considering the decay rate and the constraint on the control input. The design problem of the FJR is defined as follows:

Maximize α

$$
\begin{aligned}
& X > 0 \\
& \left[-XA_1^T - A_1X + M_1^T B^T + BM_1 - 2\alpha X \right] > 0 \\
& \left[-XA_2^T - A_2X + M_2^T B^T + BM_2 - 2\alpha X \right] > 0 \\
& \left[\begin{array}{c} -XA_1^T - A_1X - XA_2^T - A_2X + M_2^T B^T \\ + BM_2 + M_1^T B^T + BM_1 - 4\alpha X \end{array} \right] > 0 \\
& \left[\begin{array}{cc} X & M_1^T \\ M_1 & \mu^2 I \end{array} \right] > 0 \\
& \left[\begin{array}{cc} X & M_2^T \\ M_2 & \mu^2 I \end{array} \right] > 0 \\
& \left[X - \phi^2 I \right] > 0
\end{aligned}
\tag{54}
$$

Where $X = P^{-1}$, $M_i = F_i X$, $\mu = 4600$, $\phi = 1$.

Using the MATLAB LMI toolbox to solve the LMI conditions (50), we can get the positive definite matrix and a set of gains (51), that make the system stable.

$$\alpha = 0.072401$$

$$P = \begin{bmatrix} 0.7301 & 0.32486 & 0.096794 & 0.0034552 \\ 0.32486 & 0.55483 & 0.10616 & 0.010209 \\ 0.096794 & 0.10616 & 0.023049 & 0.0017139 \\ 0.0034552 & 0.010209 & 0.0017139 & 0.00023565 \end{bmatrix}$$

$$F_1 = [327.57 \quad 1745 \quad 261.86 \quad 57.475] \tag{55}$$

$$F_2 = [356.05 \quad 1739.2 \quad 259.77 \quad 57.5]$$

Figures 10 and 11 show the response of the system and control effort, respectively.

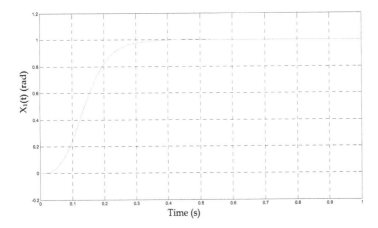

Figure 10. System responses of the single-link flexible joint, case 3.

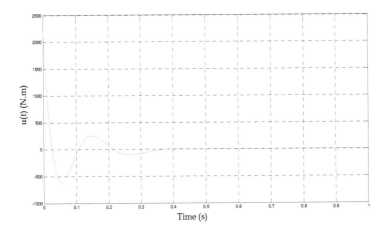

Figure 11. Control input for flexible joint robots, case 3.

Case 4: Performance-oriented parallel distributed compensation

The following stabilizing feedback gains are chosen using the pole placement method, so that K_{11} and K_{21} produce large magnitude inputs for subsystems 1 and 2, respectively, and K_{22} and K_{21} induce low magnitude inputs for those subsystems. In particular,

$$
\begin{aligned}
K_{11} &= \begin{bmatrix} 6667.2 & 4411.9 & 1052.4 & 92.6 \end{bmatrix} \\
K_{12} &= \begin{bmatrix} -33.321 & 1413.7 & 191.63 & 51.2 \end{bmatrix} \\
K_{21} &= \begin{bmatrix} 6658.7 & 4332.4 & 1025.4 & 91.1 \end{bmatrix} \\
K_{22} &= \begin{bmatrix} 72.3 & 1389.8 & 189.6 & 50.6 \end{bmatrix}
\end{aligned}
\tag{56}
$$

The required simple membership functions are selected as in Figure 12, so that, with a decrease in the corresponding plant input, in subsystems 1 and 2 respectively, the overall feedback gains come closer to K_{11} and K_{21}, and with an increase in the corresponding control input respectively, the overall feedback gains come closer to K_{21} and K_{22}. Now, the fuzzy rules for the controller are constructed as follows:

Rule 1: If $z(t)$ is $M_1(z)$ and $|u(t)|$ is "small" then $u(t) = K_{11}x(t)$

Rule 2: If $z(t)$ is $M_1(z)$ and $|u(t)|$ is "large" then $u(t) = K_{12}x(t)$

Rule 3: If $z(t)$ is $M_2(z)$ and $|u(t)|$ is "small" then $u(t) = K_{21}x(t)$

Rule 4: If $z(t)$ is $M_2(z)$ and $|u(t)|$ is "large" then $u(t) = K_{22}x(t)$

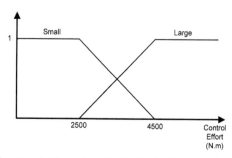

Figure 12. Membership functions for the control effort in the flexible joint robots.

A common positive definite matrix, P, satisfying the stability conditions (42) is obtained by solving the LMI problems:

$$
P = 10^4 \times \begin{bmatrix}
121710 & 15858 & 2558.5 & 63.525 \\
15858 & 8624.4 & 1458.4 & 105.36 \\
2558.5 & 1458.4 & 702.24 & 42.529 \\
63.525 & 105.36 & 42.529 & 5.0962
\end{bmatrix}
$$

Applying a unit step reference signal for $x_1(t)$, the response history and the corresponding control input are shown in Figures (13) and (14), respectively. Simulation results are investigated for the following three controllers:

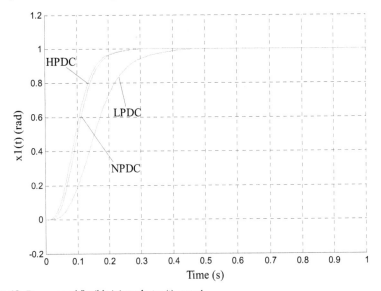

Figure 13. Response of flexible joint robot x₁(t), case 4.

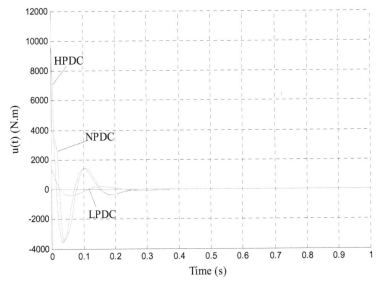

Figure 14. Control input for flexible joint robot, case 4.

1. A PDC controller with feedback gains K_{11} and K_{21} providing a high speed response, and with possible high control inputs (HPDC controller).
2. A PDC controller with feedback gains K_{22} and K_{21} providing a low speed response, and with a lower control input, as compared with the HPDC case (LPDC controller).
3. Proposed modified PDC controller, providing a fast response, yet with an acceptable level of control input (NPDC controller).

It is observed that the new controller provides a settling time similar to the HPDC case, with a much lower magnitude for the control input.

5. Conclusion

This chapter deals with approximation of the nonlinear system using Takagi-Sugeno (T-S) models with linear models as rule consequences and a construction procedure of T-S models. Also, the stability conditions and stabilizing control design of parallel distributed compensation (PDC) are discussed. It is seen that PDC a linear control method can be used to control the nonlinear system. Moreover, the stability analysis and control design problems for both continuous and discrete fuzz control systems can be transformed to linear matrix inequality (LMI) problems and they can be solved efficiently by convex programming techniques for LMIs. Design examples demonstrate the effectiveness of the LMI-based designs.

Author details

Morteza Seidi, Marzieh Hajiaghamemar and Bruce Segee
University of Maine, USA

6. References

Bonissone, P. P., Badami, V., Chaing, K.H., Khedkar, P.S., Marcelle, K. W. & Schutten, M. J. (1995). Industrial applications of fuzzy logic at general electric, *Proceedings of IEEE*, Vol. 83, No.3, (August 2002), p.p. 450-465, ISSN : 0018-9219.

Boyd, S., Ghaoui, L. E. & Feron, Eric & Balakrishnan V. (1994). *Linear Matrix Inequalities in System and Control Theory*, SIAM studies in applied mathematics, ISBN 0-89871-334X.

Chen , C. L., Chen, P. C. & Chen, C. K. (1993). Analysis and design of fuzzy control system, *Fuzzy Sets and Systems*, Vol. 57, No. 2, (July 1993) p.p. 125–140.

Ding, B. C., Sun, H. X. & Yang, P. (2006). Further study on LMI-based relaxed nonquadratic stabilization conditions for nonlinear systems in the Takagi–Sugeno's form, *Automatica*, Vol. 42, No. 3,(March 2006) p.p. 503–508.

Fang, C. H., Liu, Y. S., Kau, S. W., Hong, L. & Lee, C. H. (2006). A new LMI-based approach to relaxed quadratic stabilization of T–S fuzzy control systems, *IEEE Transactions on Fuzzy Systems*, Vol. 14, No. 3, (June 2006), p.p. 386–397, ISSN : 1063-6706.

Hong, S.K. & Langari, R. (2000). Robust fuzzy control of a magnetic bearing system subject to harmonic disturbances, *IEEE Transactions on Control System Technology*, Vol. 8, No. 2, (August 2002), p.p. 366–371, ISSN : 1063-6536.

Ichikawa, A. et al. (1993). *Control Hand Book*, Ohmu Publisher, Tokyo, in Japanese.

Joh, J., Langari, R., Jeung, E. & Chiuig, W. (1997). A new design method for continuous Takagi–Sugeno fuzzy controller with pole-placement constraints: an LMI approach, *Proceedings of IEEE International Conference on Systems, Man, and Cybernetics*, Vol. 3, p.p. 2969–2974, ISBN: 0-7803-4053-1, Orlando, Florida, USA, October 12-15, 1997.

Kawamoto, S., Tada, K., Ishigame A. & Taniguchi, T. (1992). An Approach to Stability Analysis of Second Order Fuzzy Systems, *Proceedings of First IEEE International Conference on Fuzzy Systems*, Vol. 1, pp. 1427-1434, ISBN: 0-7803-0236-2, San Diego, California, USA, March 8-12, 1992.

Kim, E., Lee, H. (2000). New approaches to relaxed quadratic stability conditions of fuzzy control systems, *IEEE Transactions on Fuzzy Systems*, Vol. 8, No. 5, p.p. 523– 534, (August 2002), ISSN : 1063-6706.

Li, J., Niemann, D. & Wang, H. O. (1998).Robust tracking for high-rise/high-speed elevators, *Proceedings of American Control Conference*,Vol.6, p.p. 3445-3449, ISBN: 0-7803-4530-4, Philadelphia, Pennsylvania, USA, June 24-26, 1998.

Lee, H.J., Park, J.B. & Joo, Y.H. (2006). Robust load–frequency control for uncertain nonlinear power systems: a fuzzy logic approach, *Information Sciences*, Vol. 176, No.23, (December 2006), p.p. 3520–3537.

Lee, K.R., Jeung, E.T. & Park, H.B. (2001). Robust fuzzy H∞ control for uncertain nonlinear systems via state feedback: an LMI approach, *International Journal of Fuzzy Sets and Systems*, Vol. 120, No.1 (May 2001) p.p. 123–134.

Lin, Y.C., & Lo, J.C. (2003). Robust H2 fuzzy control via dynamic output feedback for discrete-time systems, *Proceedings of IEEE International Conference on Fuzzy Systems*, Vol. 2, p.p. 1384–1388, ISBN: 0-7803-7810-5, May 25-28,2003.

Kang, G. & Lee, W. (1998). Design of TSK fuzzy controller based on TSK fuzzy model using pole-placement, *Proceedings of IEEE International Conference on Fuzzy Systems*, Vol. 1, p.p. 246–251, ISSN : 1098-7584, Anchorage, Alaska, USA, May 4-9, 1998.

Rhee, B. J. & Won, S. (2006). A new Lyapunov function approach for a Takagi–Sugeno fuzzy control system design. *Journal of Fuzzy Sets System*, Vol. 157, No. 9, (May 2006), p.p. 1211–1228.

Seidi, M. & Markazi, A. H. D. (2008). Model-Based Fuzzy Control of Flexible Joint Manipulator: A LMI Approach, *Proceedings of the 5th International IEEE Symposium on Mechatronics and its Applications*, ISBN: 978-1-4244-2033-9, Amman, Jordan, May 27-29, 2008.

Seidi M. & Markazi, A. H. D. (2011). Performance-oriented parallel distributed compensation ,Vol. 348, No. 7, (September 2011), pp. 1231–1244.

Spong, M. W., Khorasani, K. & Kokotovic, P. V. (1987). " An integral manifold approach to the feedback control of flexible joint robots", *IEEE Journal of Robotics and Automation*, Vol.3, No.4, (January 2003), p.p. 291-300, ISSN : 0882-4967,.

Sugeno, M. & Kang , G. T. (1986). Fuzzy Modeling and Control of Multilayer Incinerator, *Journal of Fuzzy Sets Systems*, Vol. 18, No. 3, (April 1986), p.p. 329-345.

Takagi, T., & Sugeno, M. (1985). Fuzzy identification of systems and its applications to modeling and control, *IEEE Transactions on Systems Man and Cybernetics*, Vol. 15, NO.1, (February 1985), p.p.116–132.

Tanaka, K. & Sugeno, M. (1990). Stability Analysis of Fuzzy Systems Using Lyapunov's Direct Method, *Proceedings of the North America Fuzzy Information Processing Society NAFIPS'90*, Vol. 1, pp. 133-136, Toronto, Canada, June 1990.

Tanaka, K. & Sugeno M. (1992). Stability analysis and design of fuzzy control systems, *Journal of Fuzzy Sets and Systems*, Vol.45, No.2, (January 1992), p.p.135–156.

Tanaka, K. & Sano M. (1994). A robust stabilization problem of fuzzy control systems and its applications to backing up control of a truck trailer, *IEEE Transactions on Fuzzy Systems*, Vol. 2, No. 5, (August 2002), p.p. 119-134, ISSN : 1063-6706.

Tanaka, K., Ikeda, T. & Wang, H. O. (1996). Design of fuzzy control systems based on relaxed LMI stability conditions, *Proceedings of 35th IEEE Conference on Decision and Control*, Vol. 1, , pp. 598-603, ISBN: 0-7803-3590-2, Kobe, Japan, December 11-13 1996.

Tanaka, K. & Kosaki, T. (1997). Design of a Stable Fuzzy Controller for an Articulated Vehicle, *IEEE Transactions Systems, Man and Cybernetics*, Vol. 27, No. 3, (August 2002), p.p. 552 – 558, ISSN : 1083-4419.

Tanaka, K., Ikeda, T. & Wang, H. O. (1998). Fuzzy regulator and fuzzy observer: Relaxed stability conditions and LMI-based designs, *IEEE Transactions on Fuzzy Systems*, Vol. 6, No. 2, (August 2002), p.p. 250–265, ISSN : 1063-6706.

Tanaka, K., Taniguchi, T. & Wang, H. O. (1998a). Model-Based Fuzzy Control of TORA System: Fuzzy Regulator and Fuzzy Observer Design via LMIs that Represent Decay Rate, Disturbance Rejection, Robustness, Optimality, *Proceedings of Seventh IEEE International Conference on Fuzzy Systems*, pp. 313-318, Alaska, USA, May 4-9 1998.

Tanaka, K., Ikeda, T., & Wang, H. O. (1998b). Fuzzy Regulators and Fuzzy Observers: relaxed stability conditions and LMI-based designs, *IEEE Transactions on Fuzzy Systems*, Vol. 6, No. 2, (August 2002), pp. 250-265, ISSN : 1063-6706.

Tanaka, K. &Wang, H.O. (2001). Fuzzy Control Systems Design and Analysis: A Linear Matrix Inequality Approach, 1st Edition, Wiley.

Tanaka, K., Hori, T. & Wang, H. O. (2003). A multiple Lyapunov function approach to stabilization of fuzzy control systems, *IEEE Transactions on Fuzzy Systems*, Vol. 11, No. 4, (August 2003), p.p. 582–589, ISSN : 1063-6706.

Ting, C.S. (2006). Stability analysis and design of Takagi–Sugeno fuzzy systems, *Journal of Information Sciences*, Vol. 176, No.19, (October 2006), p.p. 2817-2845.

Wang, H. O., Tanaka, K. & Griffin, M. F. (1995) "Parallel Distributed Compensation of Nonlinear Systems by Takagi-Sugeno Fuzzy Model, *Proceedings of International Joint Conference of the Fourth IEEE International Conference on Fuzzy Systems and The Second International Fuzzy Engineering Symposium*, p.p. 531-538, ISBN: 0-7803-2461-7, Yokohama, Japan, March 20-24, 1995.

Wang, H. O., Tanaka, K. & Griffin M. F. (1996). An approach to fuzzy control of nonlinear systems: Stability and the design issues. *IEEE Transactions on Fuzzy Systems*, Vol. 4, No. 1, (August 2002), p.p. 14–23, ISSN : 1063-6706.

Wang, L. X., & Mendel, J. M. (1992). Fuzzy basis functions, universal approximation, and orthogonal least-squares learning, *IEEE Transactions on Neural Networks*, Vol. 3, No. 5, (August 2002), p.p. 807–814, ISSN : 1045-9227.

Zhao, J., Wertz, V. & Gorez, R. (1996). Fuzzy gain scheduling controllers based on fuzzy models, *Proceedings of The 5th IEEE International Conference on Fuzzy Systems*, Vol. 3, No. 8, p.p. 1670–1676, ISBN: 0-7803-3645-3, New Orleans, Louisiana, USA, September 8-11, 1996.

Chapter 4

New Areas in Fuzzy Application

Muhammad M.A.S. Mahmoud

Additional information is available at the end of the chapter

1. Introduction

"The world is not black and white but only shades of gray." In 1965, Zadeh [1] wrote a seminal paper in which he introduced fuzzy sets, sets with un-sharp boundaries. These sets are considered gray areas rather than black and white in contrast to classical sets which form the basis of binary or Boolean logic. Fuzzy set theory and fuzzy logic are convenient tools for handling uncertain, imprecise, or unmodeled data in intelligent decision-making systems. It has also found many applications in the areas of information sciences and control systems.

In this chapter, we shall discuss two important categories of fuzzy logic nonlinear applications: "Control" and "Trending and Prediction". With respect to Fuzzy Control application, among the huge applications that were published under this category, two new applications are selected in this chapter to focus on the Hierarchal Control application with "multi-input" "multi-output" signals, and another application is selected as application of smart electrical grid. However, for Fuzzy Trending and Prediction Application, c-Mean Fuzzy Cluttering technique is discussed as an introduction for Fuzzy trending algorithm, and then two different applications are introduced. The first application discusses very nonlinear problem to predict the rate of accident for labours work in a construction sector, and the second application is to find a fault in complicated electrical network. All these new applications for fuzzy control and fuzzy trending recognition has been found after year 2000.

2. Control

Control is one of the main the application for the fuzzy controller, especially for the applications that can be easily expressed by linguistically. Many machines now in the market are fuzzy machines. Also the fuzzy logic has take place in the DCS's and PLC's as recognized function to build process controllers. In this chapter we shall select three applications as an example for the new application of fuzzy logic in the control.

2.1. Fuzzy control design for gas absorber system

In this section, the chapter shall present the research efforts that have been carried out on the control of gas absorbers/gas reactors. It shall also introduce the new approach to a fuzzy control design for a typical gas absorber system. The approach shall incorporate a linear state-estimation to generate the internal knowledge-base that shall store input-output pairs. This collection of pairs shall be then utilized to build a feedback fuzzy controller for the gas absorber.

2.1.1. Background

A major direction in systems engineering design has been focused on the use of simplified mathematical models to facilitate the design process. This constitutes the so-called model-based system design approach, an overview of the underlying techniques can be found in [2]. Most of the available results have thus far overlooked the operational knowledge of the dynamical system under consideration. On the other hand, a knowledge-based system approach [8] has been suggested to deal with the analysis and design problems of different classes of dynamical systems by incorporating both the simplest available model as well as the best available knowledge about the system. For single physical systems, one of the earlier efforts along this direction has been on the development of an expert learning system; see [4-7] and their references. An alternative approach has been on integrating elements of discrete event systems with differential equations [3].

A third approach has been through the use of fuzzy logic control by successfully applying fuzzy sets and systems theory [9]. In the cases where understood there is no acceptable mathematical model for the plant, fuzzy logic controllers [10] are proved very useful and effective. They are generally base on using qualitative rules of thumb, that is, qualitative control rules in terms of vague and fuzzy sentences. It has been pointed out [11] that fuzzy control systems possess the following features:

Hierarchical ordering of fuzzy rules is used to reduce the size of the inference engine. Real-time implementation, or on-line simulation, of fuzzy controllers can help reduce the burden of large-sized rule sets by fusing sensory data before imputing the system's output to the inference engine.

This section is presenting a new approach to fuzzy control design for a gas absorber system. It provides a new and efficient procedure to construct the inference engine by incorporating a linear state-estimator in generating and storing input-output pairs. This collection of pairs is then utilized to build a feedback fuzzy controller. By fine-tuning of the controller parameters, it is shown that the gas absorber system has always a guaranteed stability. Numerical simulation of a six-order gas absorber is carried out and the obtained results show clearly that the proposed estimator-fuzzy controller scheme yields excellent performance.

2.1.2. A Gas Absorber System

A. Brief Account

Separation processes play an important role in most chemical manufacturing industries. Streams from chemical reactors often contain a number of components; some of these components must be separated from the other components for sale as a final product, or for use in another manufacturing process. A common example of a separation process is gas absorption (also called gas scrubbing, or gas washing) in which a gas mixture is contacted with a liquid (the absorbent or solvent) to selectively dissolve one or more components by mass transfer from the gas to the liquid. Absorption is used to separate gas mixtures; remove impurities, contaminants, pollutants, or catalyst poisons from a gas; or recover valuable chemicals. In general, the species of interest in the gas mixture may be all components, only the component(s) not transferred, or only the component(s) transferred. Absorption is frequently conducted in trayed towers (plate columns), packed columns, spray towers, bubble columns, and centrifugal contactors. A trayed tower is a vertical, cylindrical pressure vessel in which vapor and liquid, which flow counter-currently, are contacted on a series of metal trays or plates; see Fig. 1. Components that enter the bottom of the tower is the gas feed stream are absorbed by the liquid stream, that flows across each tray, over an outlet weir and into a down-comer, so that the gas product stream (leaving the top of the tower) is more pure.

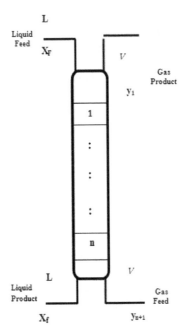

Figure 1. Gas absorption column, n stages

B. Assumptions and definitions

The basic assumptions used are:

A1) The major component of the liquid stream is inert and does not absorb into the gas stream.

A2) The major component of the gas stream is inert and does not absorb into the liquid stream.

A3) Each stage of the process is an equilibrium stage, that is, the vapour leaving a stage is in thermodynamic equilibrium with the liquid on that stage.

A4) The liquid molar holdup is constant.

We now introduce the following variable definitions:

- L = moles inert liquid per time: = liquid molar flow rate.
- V = moles inert vapor per time: = vapor molar flow rate
- M = moles liquid per stage: = liquid molar holdup per stage
- W = moles vapor per stage: = vapor molar holdup per stage
- x_j = moles solute (stage j) per mole inert liquid (stage j)
- y_j = moles solute (stage j) per mole inert vapor (stage j)

C. Dynamic model

The concept of an equilibrium stage is important for the development of a dynamic model of the absorption tower. An equilibrium stage is represented schematically in Fig. 2. The total amount of solute on stage j is the sum of the solute in the liquid phase and the gas phase (that is, $M x_j + W y_j$). Thus the rate of change of the amount of solute is $d(M x_j + W y_j)/dt$ and the component material balance around stage j can be expressed as:

$$\frac{d(M x_j + W y_j)}{dt} = L x_{j-1} + V y_{j+1} - L x_j - V y_j$$

$$\frac{dM x_j}{dt} \cong L x_{j-1} + V y_{j+1} - L x_j - V y_j \tag{1}$$

where we assumed that in accumulation, liquid is much more dense than vapor. Under assumption A4), then (1) simplifies into:

$$\frac{d x_j}{dt} \cong \frac{L}{M} x_{j-1} + \frac{V}{M} y_{j+1} - \frac{L}{M} x_j - \frac{V}{M} y_j \tag{2}$$

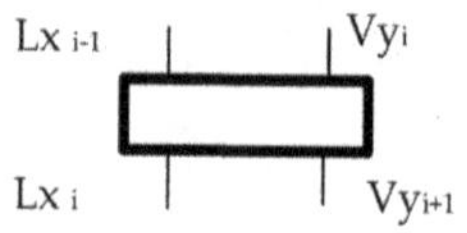

Figure 2. A typical gas absorption stage

Under assumption A3), we let

$$y_j = d \, x_j \tag{3}$$

which expresses a linear relationship between the liquid phase and gas phase compositions at stage j with d being an equilibrium parameter. Using (3) into (2) and arranging we get:

$$\frac{d \, x_j}{dt} = \frac{L}{M} x_{j-1} - \frac{(L + V \, d)}{M} x_j - \frac{V \, d}{M} x_{j+1} \tag{4}$$

For n-stage gas absorber, (4) is valid for j =2,...,n-1. At the extreme stages, we have:

$$\frac{d \, x_1}{dt} = - \frac{(L + V \, d)}{M} x_1 - \frac{V \, d}{M} x_2 + \frac{L}{M} x_f \tag{5}$$

$$\frac{d \, x_n}{dt} = - \frac{(L + V \, d)}{M} x_n + \frac{L}{M} x_{n-1} + \frac{V}{M} y_{n+1} \tag{6}$$

where xf and yn+1 are the known liquid and vapor feed compositions, respectively.

On combining (3), (4),(5) and (6), we reach the state-space model:

$$\dot{x}(t) = A \, x(t) + B \, u(t), \qquad y(t) = C \, x(t) \tag{7}$$

where A an (nxn) system matrix with a triangular structure, B is an (nxm) input matrix and C is an (nxp) output matrix given by:

$$
A = \begin{bmatrix}
-\dfrac{L+M}{M} & \dfrac{Vd}{M} & 0 & 0 & 0 & 0 & \cdots & 0 \\[2ex]
\dfrac{L}{M} & -\dfrac{L+Vd}{M} & \dfrac{Vd}{M} & 0 & 0 & & \cdots & 0 \\[2ex]
0 & \dfrac{L}{M} & -\dfrac{L+Vd}{M} & \dfrac{Vd}{M} & 0 & & \cdots & 0 \\[2ex]
0 & & \ddots & & & & & 0 \\
\vdots & & & \ddots & & & & \vdots \\
\vdots & & & & \ddots & & & \\
0 & & & & & \dfrac{L}{M} & & -\dfrac{L+Vd}{M}
\end{bmatrix} \tag{8}
$$

$$B = \begin{bmatrix} \dfrac{L}{M} & 0 \\[4pt] 0 & 0 \\ \vdots & \vdots \\ \vdots & \vdots \\ \vdots & \vdots \\ \vdots & \vdots \\ \vdots & \dfrac{V}{M} \end{bmatrix} \tag{9}$$

$$C = \begin{bmatrix} 1 & 0 & 0 & 0 & 0 & 0 \\ 0 & 0 & 0 & 0 & 0 & 1 \end{bmatrix} \tag{10}$$

2.1.3. Fuzzy controller design

The design of a fuzzy controller can be implemented by the following steps:

Step1:

Supposed that the output y (t) takes values in the interval $U = [\alpha, \beta] \subset R$. Define 2N+1 fuzzy function A^l in U that are consistent and complete with the triangular membership functions shown in Fig. 3. That is, we use the N fuzzy sets A^1, ---, A^N to cover the negative interval $[\alpha, 0)$, the other N fuzzy sets A^{N+2},---, A^{2N+1} to cover the positive interval $(0, \beta]$, and choose the center x N+1 of fuzzy set A^{N+1} at zero.

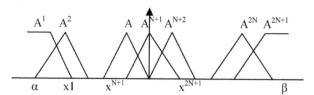

Figure 3. Membership functions for the fuzzy controller.

Step 2:

Consider the following 2N+1 fuzzy IF-THEN rules:

$$\text{IF } y \text{ is } A^l, \text{ THEN } u \text{ is } B^l \tag{11}$$

Where l = 1, 2, ---, 2N+1, and centers $\overline{y}^l$ of fuzzy set B^l are chosen such that,

$$\overline{y}^l \begin{cases} \leq 0 & \text{for } l=1,\cdots,N \\ = 0 & \text{for } l=N+1 \\ \geq 0 & \text{for } l=N+2,\cdots,2N+1 \end{cases} \tag{12}$$

Step 3:

Design the fuzzy controller from the 2N+1 fuzzy IF THEN rules (11) using product inference engine, singleton fuzzifier and center average defuzzifier; that is, the designed fuzzy controller is

$$v=-f(y)=\frac{\sum_{1=1}^{2N+1}\overline{y^{1}}\mu A^{1}(y)}{\sum_{1=1}^{2N+1}\mu A^{1}(y)} \tag{13}$$

Where $\mu_{A^l}(y)$ are shown in Fig. 3 and y^1 satisfy $\overline{y}$ (12).

To estimate the range of the input-output pairs $\{v_i, y_i\}$, full order estimator [2] can be used.

2.1.4. Simulation studies

Consider a gas absorber system with the following parameters: L=80, M=200, V=100 and d=0.5.

Thus,

$$\frac{L+Vd}{M}=-0.65,\frac{L}{M}=0.4,\frac{V\,d}{M}=0.25,\frac{V}{M}=0.5$$

A MATLAB program is written to simulate the gas absorber system. Different positive and negative step input are applied to estimate the outputs. The results of two cases are illustrated in Fig. 4 and Fig. 5. The tracking behaviour of the outputs is shown.

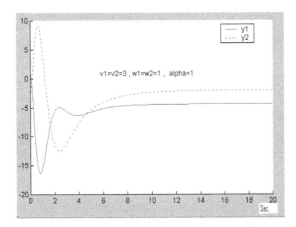

Figure 4. Output response with positive step input signal

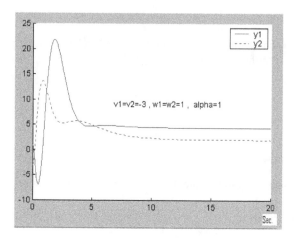

Figure 5. Output response with negative step input signal

From the input-output pair obtained, the behaviour of the system is examined and the ranges of its outputs (controllers' inputs) are predicted. Fig. 6 illustrates a block diagram of the gas absorber and the fuzzy controller array.

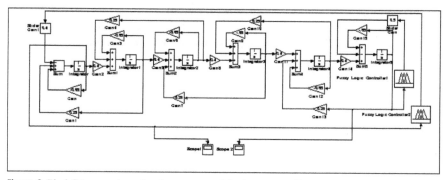

Figure 6. Block Diagram of gas absorber system and the fuzzy controllers

To control the response of the gas absorber, the range of linguistic values of the output of each feedback fuzzy controller is tuned between (– 3) and (3). Comparison between the output response with fuzzy controller (when the number of linguistic values of the controller input – output pair is three) and without controller is illustrated in Fig.7 and Fig.8.

In Fig.7, the controller is tuned to interfere the natural decay of the system. In Fig. 8, the fuzzy controller is adjusted to improve the response of the gas absorber. It is noted that the response of controlled system has less overshoot, less steady state error and faster compared to the uncontrolled system.

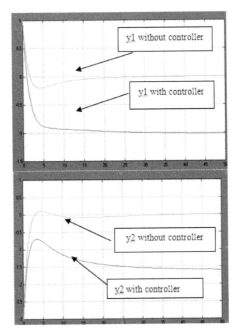

Figure 7. Controller is tuned to interfere the natural decay of the system

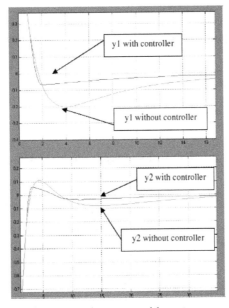

Figure 8. The controller is tuned to improve the response of the system.

2.1.5. Discussion

This section has presented a new and simple fuzzy controller for a gas absorber system to enhance the response of the output. The simulation results have shown that the controller guarantees well-damped behaviour of the controlled gas absorber system.

2.2. Large scale fuzzy controller

In this section, we shall develop a new approach to the control of interconnected system using fuzzy system theory. The approach shall be based on incorporating a group of local estimators on the system level to generate the input-output database. An array of feedback fuzzy controllers shall then be designed to ensure the asymptotic stability of the closed loop system. The developed technique shall be applied to an unstable large-scale system and extensive simulation studies shall be carried out to illustrate the potential of this new approach..

2.2.1. Background

In control engineering research, problems of decentralized control and stabilization of interconnected systems are receiving considerable interest in recent years [14,15] where most of the effort is focused on dealing with the interaction patterns. It is concluded that a systematic approach to deal with the problems of interconnected systems is twofold: first is to base the analysis and design effort on the subsystem level using conventional control methods and second is to deal with interactions effectively. These methods are facilitated, in general, by virtue of several mathematical tools including linearization, delay approximation, decomposition and model reduction. This constitutes the so-called model-based control system approach for which we have seen numerous techniques [16]. Most of the available results have so far overlooked the operational knowledge of the interconnected system under consideration. In [17], a knowledge-based control system approach has been suggested to deal with the analysis and design problems of interconnected systems by incorporating both the simplest available model as well as the best available knowledge about the system. For single physical systems, one of the earlier efforts along this direction has been on the development of an expert learning system [18-19]. An alternative approach has been on integrating elements of discrete event systems with differential equations [20]. A practically-supported third approach has been through the use of fuzzy logic control by successfully applying fuzzy sets and systems theory [21].

For interconnected systems, the foregoing approach motivates the research into intelligent control by combining techniques of control and systems theory with those from artificial intelligence. The main focus should be on integrating a knowledge base, an approximate (humanlike) reasoning and/or a learning process within a hierarchical structure.

Fuzzy logic controllers [23-25] are generally considered applicable to plants that are mathematically poorly understood (there is no acceptable mathematical model for the plant) and where experienced human operators are available for satisfactorily controlling the plant and providing qualitative "rules of thumb" (qualitative control rules in terms of vague and fuzzy sentences).

A concerted effort has been made to formally reduce the size of the fuzzy rule base to make fuzzy control attractive to interconnected systems. Two of the difficulties with the design of any fuzzy control system are:

- The shape of the membership functions.
- The choice of fuzzy rules.

The properties that a fuzzy membership function is used to characterize are usually fuzzy. Therefore, we may use different membership functions to characterize the same description.

Conceptually, there are two approaches to determine a membership function. The first approach is to use the knowledge of human experts. Usually this approach can only give a rough formula of the membership function; fine-tuning is required. In the second approach, data are collected from various sensors to determine the member ship functions. Specifically, the structures of the membership functions are specified first, then fine-tuning of the membership function parameters should be implemented based on the collected data [8].

In this section, we contribute to the further development of intelligent control techniques of interconnected systems. It provides a new approach to fuzzy control design for interconnected system. The approach consists of two stages: In the first stage, a group of local state estimator is constructed to generate the data base of input-output pairs. In the second stage, an array of feedback fuzzy controllers is designed and implemented to ensure the asymptotic satiability of the interconnected system. Simulation studies on a large-scale system with unstable eigenvalus are carried out to illustrate the features and capability of this new approach.

2.2.2. State estimation of interconnected systems

In the sequel, the terms large-scale and interconnected are used interchangeable. The term large scale system (LSS) does not have a unique established meaning, but it covers systems that possess several particular feature, such as multiple subsystem, [14,17] multiple control, multiple objectives, decentralized and/or hierarchical information structures. Any LSS includes many variables but their control is faced by a well-know fact [16] that the states are not always available for measurement and state must be estimated.

Many authors have considered the state estimation of large-scale systems in input decentralized fashion. Here we summarize one convenient algorithm [15]. Let the state model of the ith subsystem described by

$$\dot{x}_i(t) = A_i x_i(t) + B_i u_i(t) + \sum^N G_{ij} x_j \qquad (14)$$

$$y_i(t) = C_i x_i(t), \ i,j = 1, 2, \ldots\ldots N \qquad (15)$$

Where all vectors and matrices are appropriately defined and $g_i(.)$ is the interaction function between the ith subsystem and the rest of the system. It is considered that (C_i, A_i) is completely observable for i = 1, 2, ... N.

The following algorithm finds the optimal states of a large-scale system based on decentralized estimation and control [17]:

Algorithm 1:

Step 1:

Read the matrices A_i, B_i and select $Q_i \geq 0$ and $R_i > 0$ as weighted matrix.

Step 2:

Solve the following 2N algebraic Riccati equations for H_i, K_i

$$Hi(A^T_i + \alpha I_i) + (A_i + \alpha I_i)H_i - H_i D_i H_i + Q_i = 0 \tag{16}$$

$$Ki(A^T_i + \alpha I_i) + (A_i + \alpha I_i)K_i - K_i S_i K_i + Q_i = 0 \tag{17}$$

Where $D_i = C^T_i C_i$, $S_i = B_i R^{-1}_i B^T$

Step 3:

Integrate the following set of N simultaneous equation for $e_i(t)$, I = 1, 2 .... N, using the initial condition $e_i(0) = x_i(0)$

$$\begin{pmatrix} \dot{e}_1 \\ \vdots \\ \dot{e}_N \end{pmatrix} = \begin{pmatrix} A_1 - S_1 K_1 & \cdots & G_{1N} \\ \vdots & \ddots & \\ G_{N_1} & & A_N - H_N D_N \end{pmatrix} \begin{pmatrix} e_1 \\ \vdots \\ e_N \end{pmatrix} + \begin{pmatrix} B_1 v_1 \\ \vdots \\ B_N v_N \end{pmatrix} \tag{18}$$

Step 4:

Integrate the following set of n simultaneous equations for x 1(t), i = 1, 2, ... N

$$\begin{pmatrix} \dot{x}_1 \\ \vdots \\ \dot{x}_N \end{pmatrix} = \begin{pmatrix} A_1 - S_1 K_1 & \cdots & G_{1N} & S_1 K_1 \cdots & 0 \\ \vdots & & & & \ddots \\ G_{N_1} & & A_N - S_N I_N & 0 & S_N K_N \end{pmatrix} \begin{pmatrix} x_1 \\ \vdots \\ x_N \end{pmatrix} + \begin{pmatrix} B_1 v_1 \\ \vdots \\ B_N v_N \end{pmatrix} \tag{19}$$

$\dot{x}$ N G_{N1} A_N - S_N I_N 0 S_N K_N x N B_N v_N

Step 5:

Generate the input-output pairs $\{v_i, \hat{y}_i = c_i \hat{x}_i\}$.

2.2.3. Interconnected system

Assume the following interconnected system of order 10 [17]:

$$
A=\begin{pmatrix}
-1.5 & -0.3 & -0.25 & 0.1 & 0.5 & rl_{11} & rl_{12} & rl_{13} & rl_{14} & rl_{15} \\
0.1 & 0 & 0 & -0.2 & 0 & rl_{21} & rl_{22} & rl_{23} & rl_{24} & rl_{25} \\
0 & 0.2 & -1 & 0 & 0.4 & rl_{31} & rl_{32} & rl_{33} & rl_{34} & rl_{35} \\
0.6 & -0.1 & -0.25 & -2 & 0 & rl_{41} & rl_{42} & rl_{43} & rl_{44} & rl_{45} \\
0.4 & 0.2 & 1 & 0.5 & 0.1 & rl_{51} & rl_{52} & rl_{53} & rl_{54} & rl_{55} \\
r2_{11} & r2_{12} & r2_{13} & r2_{14} & r2_{15} & -1.5 & -0.3 & -0.25 & 0.1 & 0.5 \\
r2_{21} & r2_{22} & r2_{23} & r2_{24} & r2_{25} & 0.1 & 0 & 0 & -0.2 & 0 \\
r2_{31} & r2_{32} & r2_{33} & r2_{34} & r2_{35} & 0 & 0.2 & -1 & 0 & 0.4 \\
r2_{41} & r2_{42} & r2_{43} & r2_{44} & r2_{45} & 0.6 & -0.1 & -0.25 & -2 & 0 \\
r2_{51} & r2_{52} & r2_{53} & r2_{54} & r2_{55} & 0.4 & 0.2 & 1 & 0.5 & 0.1
\end{pmatrix} \qquad (20)
$$

$$
B=\begin{pmatrix}
0 & 0 & 0 & 0 \\
1 & 0 & 0 & 0 \\
0 & 0 & 0 & 0 \\
0 & 1 & 0 & 0 \\
0 & 1 & 0 & 0 \\
0 & 0 & 0 & 0 \\
0 & 0 & 1 & 0 \\
0 & 0 & 0 & 0 \\
0 & 0 & 0 & 1 \\
0 & 0 & 0 & 1
\end{pmatrix} \qquad (21)
$$

$$
C=\begin{pmatrix}
1 & 1 & 0 & 0 & 0 & 0 & 0 & 0 & 0 & 0 \\
0 & 0 & 1 & 0 & 1 & 0 & 0 & 0 & 0 & 0 \\
0 & 0 & 0 & 0 & 0 & 1 & 1 & 0 & 0 & 0 \\
0 & 0 & 0 & 0 & 0 & 0 & 0 & 1 & 0 & 1
\end{pmatrix} \qquad (22)
$$

Which is considered to be composed of two-coupled subsystems; each of order 5. The coupling parameters are rl_{jk} and $r2_{jk}$ where j and k take values of 1,2,3,4 and 5. In the sequel, we refer to the structure of the interconnected system model as:

$$
\dot{x}=\begin{pmatrix}
A11 & \cdots & G12(\underline{r1}) \\
\vdots & \ddots & \\
G21(\underline{r2}) & & A22
\end{pmatrix} x + \begin{pmatrix} B1 \\ \vdots \\ B2 \end{pmatrix} v \qquad (23)
$$

Where G12(r1) and G21(r2) are the coupling mtrices.

For a typical values [4] of $rl_{15}=-0.1$, $rl_{24}=0.1$, $rl_{42}=0.2$, $r2_{22}=0.1$, $r2_{42}=0.15$, $r2_{51}=0.11$ and allothe values of coupling parameters are zeros, we examined the stability of the system by computing the eigenvalues of matrix A. They are {-1.0915, -1.0641, 0.477 + j0.0206, 0.477 – j0.00206, 0.022 + j0.0544, 0.022 – j0.0544, -1.8709 + j0.1713, -1.8709 – j0.1713, -1.9306 + j0.1413, -

1.9306 – j0.1413}, and it is quite clear that there are four eigenvalues lie in the open right half of the complex plane and thus the interconnected system is unstable. Further, it is easy to check that the interconnected system is both controllable and observable.

2.2.4. Estimation of the system state variables and outputs

A Matlab program is written to implement the computational Algorithm (1) of section 1.2.2 on the interconnected system. Different positive and negative step input are applied to estimate the outputs. The results of two cases are illustrated in Fig. 9 and Fig. 10 It is observed that the outputs tend to track conveniently the input signals.

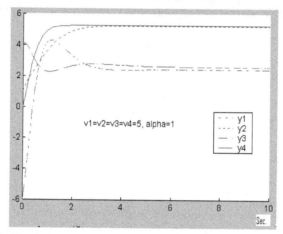

Figure 9. Simulation Results for case 1.

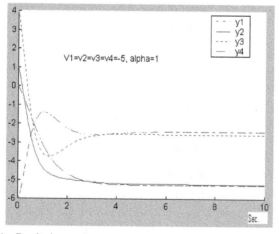

Figure 10. Simulation Results for case 2.

2.2.5. Design of an array of fuzzy controller

We are going to treat the interconnected system at hand as being composed of two identical and coupled subsystems. The control system to be designed is such that each subsystem has its own fuzzy negative feedback controller which its input being the output of the respective subsystem (Fig. 11). Each subsystem fuzzy controller is constructed using two fuzzy systems.

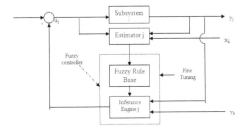

Figure 11. Block diagram of the proposed fuzzy feedback controller array.

In order to build each fuzzy controller, the following steps are implemented:

Step 1:

The range of the inputs to each fuzzy controller $[\alpha_i, \beta_i]$ are driven from the estimated value of the respective subsystem outputs, where i = 1, 2, 3, 4.

Step 2:

2N+1 fuzzy set M^L_i in $[\alpha_i, \beta_i]$ that are normal, consistent and complete with triangular membership functions [24], are defined for each controller, where L = 1, 2--, 2Ni+1. That is we use N_i fuzzy set M^1_i, ---, $M_i^{N_i}$ to cover the negative internal $[\alpha_I, 0)$, the other N_i fuzzy sets $M_i^{N_i+2}$, ---, $M_i^{2N_i+1}$ to cover the positive internal $(0, \beta]$, and the center of fuzzy set $M_i^{N_i+1}$ at zero.

Step 3:

The following 2Ni+1 rules are considered

$$\text{IF } y_{ai} \text{ is } M_i^L \text{ or } y_{bi} \text{ is } M_i^L \text{ then u is } K_i^L$$

Where L = 1, 2,---, 2Ni+1, and a_i, b_i are the input to the fuzzy controller i, and the center y_{ai}^L and y_b^{-L} of the fuzzy set k_i^L are chosen such that

$$y_{ai}^{-L} \text{ and } y_{bi}^{-L} \begin{cases} \leq -0 & \text{for } 1=1,\cdots,N_i \\ =0 & \text{for } 1=N_i+1 \\ \geq 0 & \text{for } 1=N_i+2,\cdots,2N_i+1 \end{cases} \quad (24)$$

Step 4:

Product inference engine, singleton fuzzyfier, and center average defuzzifier are selected to design the fuzzy controller.

2.2.6. Simulation results

The behaviour of the interconnected system outputs after implementing the fuzzy controllers with unity step function input are shown in Fig 12 and Fig: 13. It is clearly evident that the system becomes asymptotically stable by using the negative fuzzy feedback controller array.

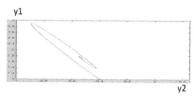

Figure 12. Outputs y1 against y2 y2

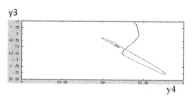

Figure 13. Outputs y3 against y4 y4

2.2.7. Performance of the proposed fuzzy feedback controller array

Now, we examine the effect of coupling matrices on the performance of fuzzy controlled interconnected system. Five additional cases with deferent coupling ranks are implemented. Fine tuning of membership functions was required to adjust their ranges. The following table summarizes the test cases:

Case No.	A11,A2 2Norm	G12 Sparsty	G12 Norm	G21 Sparsty	G21 Norm	System Stability without controller	System Stability with controller
1 (Fig 12, 13)	2.2529	3/25	0.2	3/25	1.8028	Unstable	Stable
2 (Fig. 14, 15)	2.2529	12/25	0.4712	3/25	0.1803	Unstable	Stable
3 (Fig. 16, 17)	2.2529	3/25	.2	12/25	0.5341	Unstable	Stable
4(Fig. 18, 19)	2.2529	1	3.0361	3/25	0.1803	Unstable	Stable
5 (Fig. 20, 21)	2.2529	3/25	.2	1	3.0364	Unstable	Stable
6 (Fig.22, 23)	2.2529	1	3.0361	1	3.0417	Unstable	Stable

Table 1. Results summary for 6 test cases.

The following figures illustrate the above test cases:

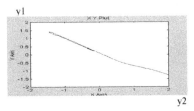

Figure 14. Case 2 Outputs y1 against y2

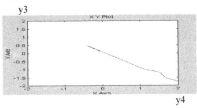

Figure 15. Case 2 Outputs y3 against y4

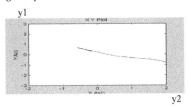

Figure 16. Case 3 Outputs y1 against y2

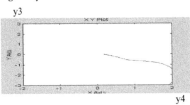

Figure 17. Case 3 Outputs y3 against y4

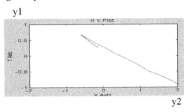

Figure 18. Case 4 Outputs y1 against y2

y3
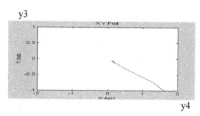
y4

Figure 19. Case 4 Outputs y3 against y4

y1
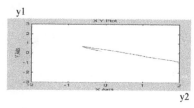
y2

Figure 20. Case 5 Outputs y1 against y2

y3
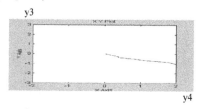
y4

Figure 21. Case 5 Outputs y3 against y4

y1
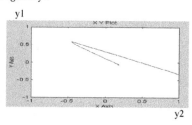
y2

Figure 22. ase 6 Outputs y1 against y2

y3
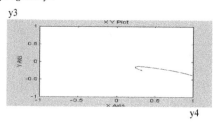
y4

Figure 23. Case 6 Outputs y3 against y4

2.2.8. Discussion

This section has developed a new fuzzy control design approach to interconnected system. It has been shown the approach consists of two stages: In stage 1, a group of local state estimator has been constructed to generate the input-output database. Then an array of feedback controllers has been designed and implemented to guarantee the overall asymptotically system stability. Extensive simulation studies have been performed to support the developed design approach.

2.3. Power factor correction

This section presents the use of fuzzy logic technique to control the reactive power of a load and hence improve the power factor. A shunt compensator is used, which consists of a reactor in series with a phase controlled Thyristor bridge in parallel with a capacitor. The control composed of two independent fuzzy controllers, the Fuzzy Grouse Controller (FGC) and the Fuzzy Fine Controller (FFC). These fuzzy controllers are used to control the firing angle of the Thyristor Bridge until the source power factor reaches a desired value. Simulations for three different practical study cases are presented and the results show how the designed controller is fast and accurate.

2.3.1. Background

Power factor, nowadays, is an important issue. The over increasing utilization of power electronics in all kinds of industry applications and the severe standards requirements are pushing the research toward new solutions to keep the industrial power factor within certain ranges [26].

A mathematical formulation for the optimal reactive power control is discussed in [27]. Also, the optimized Fuzzy logic and digital PID controllers for a single phase power factor correction converter used in online UPS are demonstrated in [28, 29]. Parameters such as input membership functions, output membership functions, inference rules of fuzzy logic controller and proportional gain, integral gain and derivative PID controller are selected and optimized by genetic algorithms. In additional to that, the applications of a hybrid converter are implemented in [30]. The hybrid converter is basically a converter bridge with two GTOs. A control strategy based on learning is proposed. The learning structure is coded into Fuzzy conditional rules to train a neural network in manipulating the converter variables.

The common method of correction is by means of using static capacitors, whether connected in series or parallel. These are installed as a single unit or as a bank, to regulate the voltage and the reactive power flow at the point of connection. In shunt compensation arrangement, a reactor is connected in parallel with conventional capacitor compensation. The shunt reactor current can be varied via a phase-controlled thyristor bridge connected in series with the shunt reactor [31]. Changing the thyristor firing angle varies the amount. of the current flowing through the reactor. Thus, this thyristor controlled reactor acts as a variable reactor

[32]. By varying the firing angle, the total reactive power of the system can be controlled and hence the power factor of the system is improved. The use of fuzzy logic to derive a practical control scheme for a boost rectifier with reactive power factor correction was applied [33]. The control action is primarily derived from a set of linguistic rules used to generate a slow-varying DC signal to determine the PWM ramp function. The proposed technique uses lesser sensing elements than the classical rectifier. A new FACTS controller known as the Bootstrap variable inductance can emulate a variable positive and negative inductance [34]. The bootstrap variable inductance has a variety of FACTS applications such as series compensation of lines, fault current limiting, reactive-power control and load power factor improvement.

Here in this section, we shall focuses on the use of fuzzy logic sets to control the supply power factor. Unlike the conventional capacitive approach for power factor improvement in ac power system, the proposed control scheme has the advantage to avoid complexities associated with the non-linear mathematical modelling of switching converters. The proposed fuzzy logic controlling scheme consists of two controllers. The first controller (FGC) is designed to give the nearest desired value of the firing angle required to compensate for the source reactive power. However, the output correction of this controller is not efficiently accurate and hence, another correction step is needed. Thus, the second controller (FFC) checks the value of the source power factor and improves it above a pre-set desired value. The discussion includes the following:

- Illustration of the proposed control scheme.
- Description of the design steps of the power factor controller.
- Simulation of the proposed technique by testing it for three different study cases.

2.3.2. Fuzzy power factor controller

Figure 24 illustrates the block diagram for a single-phase variable load, with variable lag power factor, supplied by sinusoidal AC power source. Capacitor bank in parallel with inductance, controlled by single-phase full-wave circuit, are connected in parallel to the load in order to govern the total reactive power of the circuit. Fuzzy controller is designed to tune the firing angle of the single-phase full-wave circuit in order to adjust the voltage applied across the Inductance. By this way, the total source reactive power can be minimized to improve the source power factor. For three-phase circuits, one controller is dedicated for each phase. Here, we considered single-phase circuit for simplicity.

The structure of the controller contains two independent fuzzy controllers: Fuzzy Grouse Controller (FGC) and Fuzzy Fine Controller (FFC). FGC input is the load reactive power, the output of this controller gives the nearest value of the desired firing angle, which required to minimize the source power factor. FFC input is the source power factor. The output of FFC corrects the firing angle of the single-phase full-wave controller until the source power factor reaches or exceeds the pre-setted desired value.

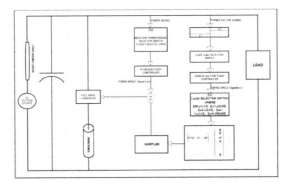

Figure 24. A detailed block diagram of the load, the fuzzy controllers, the source and the control scheme

The following procedures describe the steps of designing the power factor controller:

2.3.2.1. Elements sizing

The sizes of the 'Inductance' and the capacitance bank are selected such that their maximum available reactive power (in VAR) is equal to the maximum load reactive power (MLQ). Since the full source voltage is continuously applied on the capacitance bank (assuming the voltage drop across the short cable is negligible), then capacitance value 'C' of bank can be determined as follows:

$$C = \frac{MLQ}{2 \times \pi \times f \times V_{source}^2} \quad Farad \tag{25}$$

Where V_{source} is "source" r.m.s. voltage in Volt and f is "source" frequency in Hz.

However, this is not the case for the inductance since it is connected in series with a full wave controller. The existence of such controller will limit the available maximum reactive consumed by the inductance depending on the firing angle action of the thyristor bridge. Thus, the effect of the single-phase full-wave circuit is considered when determining the size of the inductance.

As listed in [35], the general formulas for the r.m.s value of current (I_{load}) and voltage (V_{load}) across a load, comprises of inductance in series with resistance, controlled by single-phase full-wave circuit. These formulas are given as follows:

$$I_{Load} = \frac{\sqrt{2} \times V_{source}}{z} \left[\frac{1}{\pi} \int_{\pi}^{\beta} \{ \sin(\omega t - \theta) - \sin(\alpha - \theta) e^{(r/l)(\alpha/\omega - t)} \} d\omega t \right]^{1/2} \tag{26}$$

$$V_{Load} = \frac{\sqrt{2} \times V_{source}}{z} \left[\frac{1}{\pi} \int_{\pi}^{\beta} \{ \sin(\omega t - \theta) - \sin(\alpha - \theta) e^{(r/l)(\alpha/\omega - t)} \} d\omega t \right]^{1/2} \tag{27}$$

Where:

$\omega = 2\pi f$ radian/second
α = Firing angle
β = Extinction angle (cut-off angle)
$\theta = \tan^{-1}(l/r)$
l = Load Inductance
r = Load Resistance
t = Time
z = Load impedance

Since the conducting angle $\delta = \beta - \alpha$ cannot exceed π, the firing angle α may not be less than θ and the control range of the firing angle is:

$$\pi \geq \alpha \geq \theta \tag{28}$$

For maximum available reactive power consumed by pure inductance MARP where r is ignored in the above equation 24 & 25), the maximum conducting angle is considered. Therefore, the value of the inductance can be calculated from the following equation assuming that $\alpha = \pi/2$ and $\beta = \pi$ (neglecting the impedance of the short Cable and assuming ideal thyristors):

$$\text{MARP} = \left(V_{load}\right) \times \left(I_{load}\right) \tag{29}$$

This MARP value must be equal to the maximum load reactive power MLQ for complete compensation. However, the two equations listed above are nonlinear and difficult to solve. Another simple procedure is needed when determining the size of the inductance L.

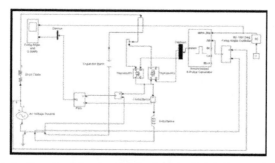

Figure 25. Simulink circuit used to determine and plot the ratio (QL/MLQ) verses firing angle (α)in the range from $\pi/2$ to π degree.

Thus, the value of the inductance can be found by a practical and fast method using Simulink tool [26]. Figure 25, illustrates the proposed model that been used. The model consists of an ac power source connected to a reactive power compensator controlled by a thyristor circuit. The procedure to find the value of the inductance L can be summarized by the following steps:

- Select an initial value for the inductance as $L = 1 (\omega^2 C)$
- Run the model and record the source VAR.
- If the source VAR equals zero, then stop the trials. If not, increase slightly the value of the inductance until the source VAR reaches the zero value. Then, this value of the source VAR will determine the desired value of the inductance.

2.3.2.2. Fuzzy Grouse Controller (FGC) design

After using the Simulink model shown in Figure 25, the inductance value is obtained from the iteration. Then, the same model is used again to find the ratio of (Q_L/MLQ) for a range of (α) starting from $\pi/2$ to π degree where (Q_L) is the Inductance reactive power at certain value of (π). This ratio (Q_L/MLQ) verses the firing angle (α) is then plotted. Typical curve is shown in Figure 26.

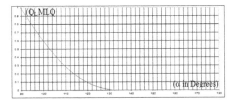

Figure 26. Typical curve for (Q_L/MLQ) verses firing angle (α)

After that, the resultant nonlinear curve is divided to N sections where each nonlinear section is approximated by the nearest linear section. Thus, N Gross Fuzzy controllers (FGC) are built, one FGC for each section, such that the input of each controller is equal to 1-(Q_{Load}/(MLQ), where (Q_{Load}) is the load reactive power. The output of the FGC is the firing angle (α), which results in (Q_L) approximately equal to MLQ minus Q_{Load}. The accuracy of the resultant (Q_L) depends on the curve linearization.

In order to design each FGC, let the full range of the input fuzzy membership function FGCMF(in)[a] for each controller is set to the (Q_L/MLQ) limits of the respective linearized section, and the range of the output fuzzy membership function FGCMF(out)[b] for each controller is set to the (α) limits of the respective linearized section. For example, in Case 3 (as will be explained later into details), the first controller is designed to take an action in case of (QL/MLQ) ratio reaches a value between rl=0.95 and 1and, the output firing angle of this controller shall take a value between 90 degree and α_1 = 100 degree. However, the second controller is designed to take an action in case of (QL/MLQ) ratio reaches a value between r2 = 0.175 and rl = 0.950 and, the output firing angle of this controller shall take a value between α_1 = 100 degree and α_2 = 105degree, and so on.

The resultant output of each fuzzy controller can be obtained based on the respective linearized section using the following functions for fuzzy implication process:

- Mamdany engine.
- Triangle type for Membership functions.
- Product for And.
- Max for Or.

- Proportional for Aggregation.
- Largest of Maximum for Defuzzification.
- Fuzzy rule : IF is MF(in)[a] THEN (α) is MF(out)[b]

where,

a = 1,2, ... F (fuzzy membership function number)
b = F-a+l(fuzzy membership function number)
a and b are fuzzy membership function numbers
F is the number of the fuzzy membership function.

Simulink automatic switching system SS 1 as shown in Figure 24 is designed to check the active range of (Q_L/MLQ) in order to select the proper FGC based on that value of (Q_L/MLQ) ratio.

2.3.2.3. Fuzzy Fine Controller (FFC) design

Since the FGC output is not accurate due to the linearization process, FFC is designed to tune the firing g angle in order to achieve the desired power factor.

Two fuzzy controllers are used along with automatic switching system to select the proper controller based on the power factor type (Lead or Lag), which is determined from the source VAR sign (+ or -) as shown in Figure 24.

The input of each FFC controller is the source power factor (PF) and the output is the corrective firing angle (M), which is required to fine tune the FGC output. Since the power factor value varies between zero and one, then the range of each FPC input membership function FFCMF(in)[a] is [0, 1].

The output membership function of each FFC is FFCMF(out-Lag)[a] for lagging input and FFCMF(out-Lead)[a] for leading input and, the magnitude of the firing angle range for the first linearzed section (α_1) is considered as a base to scale the FFC output as given hereinafter with the fact that any other section can be selected as the base. Also, [0, $\Delta\alpha_1$] and [-$\Delta\alpha_1$, 0] are assigned to FFCMF(out-Lag)[b] and FFCMF(out-Lead)[b] respectively. In addition to that, N multiplier factors ($k_N = \Delta\alpha_N / \alpha_1$) are used to scale the FFC output in order to match the magnitude of the firing angle range of the respective linearzed section ($\Delta\alpha_N$). By this method, less number of FFC's are used.

Another Simulink automatic switching system SS2 as shown in Figure 24 synchronized with SS 1 is designed to check the active range of (Q_L/MLQ) in order to select the proper multiplier factor (K_N). The fuzzy implication process functions used for FGC design are used for each FFC design as well, but with the following fuzzy rules:

$$\text{IF } (PF) \text{is FFCMF} (in)^a \text{THEN}(\alpha_\Delta) \text{ is FFCMF} (out\ Lag)^a \qquad (30)$$

$$\text{IF } (PF) \text{is FFCMF} (in)^a \text{THEN}(\alpha_\Delta) \text{ is FFCMF} (out\ Lead)^a \qquad (31)$$

2.3.2.4. Discrete control signal design

Simulink Sum Block is used to add the output of GFC to the output of FFC. Sampler system shown in Figure 24 is designed to convert the resultant analog control signal to discrete signal. The sampling time is selected to be greater than the system time constant. The discrete signal is connected to the input for the synchronized pulse generator to control the thyristor firing angle. Accordingly, the network var and the source power factor are controlled.

2.3.3. Case study

The Simulink circuit shown in Figure 24 is used as a base to study three 'test' cases. These cases are assigned to check the capability of the controller to operate within a considerable variation of power factor values at different loading and voltage level.

			Case 1	Case 2	Case 3
Source Voltage (Volt)			120	480	4160
Load Stages	Stage 1	P1 (kW)	2	40	40
		Q1(kVAR)	3.197	142.857	142.857
	Stage 2	P2 (kW)	12	200	240
		Q2 (kVAR)	12.789	571.429	428.571
	Stage 3	P3(kW)	32	1200	1240
		Q3 (kVAR)	22.38	1000	1000
	Stage 4	P4 (kW)	42	2240	2280
		Q4 (kVAR)	0	0	0
MLQ (MVAR)			0.02238	1	1
F (Number of fuzzy membership functions)			7	7	7
k1			1	1	1
k2			2.8	0.95	0.5
k3			0.44	1	2.3
k4			5.7	4.5	5.3
r1			0.64	0.52	0.95
r2			0.28	0.21	0.2
r3			0.03	0.02	0.04
ALPHA1 (DEGREE)			99	102	100
ALPHA2 (DEGREE)			124	113.5	105
ALPHA3 (DEGREE)			128	125.5	128
N (Number of sections)			4	4	4

Table 2. Circuit data and parameters for the test cases

2.3.4. Test cases data

Each test case consists of four load stages. The load stages are selected such that the power factor varies from 0.3 to 1.0 lag. However, for practical cases, the power factor varies between 0.6 to 0.8 and thus the proposed wide range of power factor tested here is to demonstrate the capability of the designed controller. Table 2 summaries the circuit parameters for three voltage levels 120. 480 and 4160 Volts respectively. Figures 27-29 illustrate the linearization process for each case.

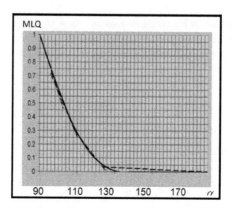

Figure 27. Linearization results for case1.

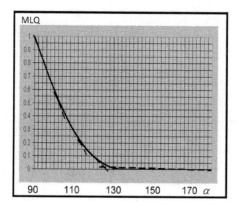

Figure 28. Linearization results for case 2

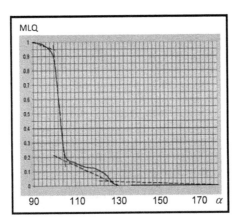

Figure 29. Linearization results for case 3

2.3.5. Results

The controller is adjusted to correct the power factor of the test cases to a value greater than a desired value of 0.97. This value is the pre-set value and it can be any chosen practical value. Figures 30 -32 illustrate the results for cases 1-3 respectively. These figures show the variation of the load active and reactive power with respect to time for each case. The response of the controller during the test period represented by the firing angle is also shown. In addition to that, source and load power factor values are plotted to check the response time of the controller and its accuracy.

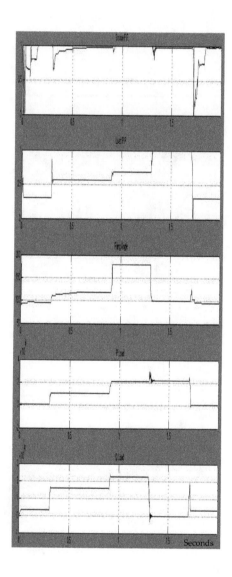

Figure 30. Results of test case 1.

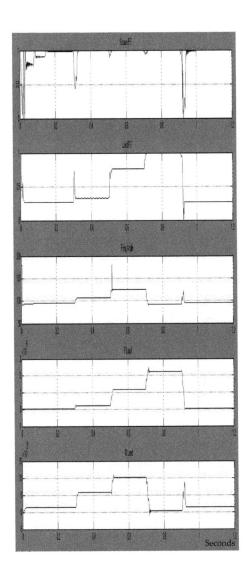

Figure 31. Results of test case 2.

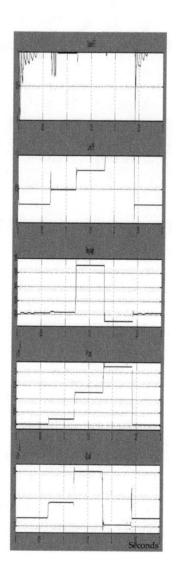

Figure 32. Results of test case 3.

2.3.6. Discussion

The test results for the three cases show clearly how efficient is the controller. Even when the load reactive power is very small at both high and low power factor, the controller was successful in reaching an accurate level. During the stage where the load power factor is greater than 0.97 and, hence no need for capacitor compensation, the controller will check the source power factor at the beginning of that stage and if it drops below 0. 97, it will take an action in order to eliminate the compensation added in the previous stage. That is why the controller took an action as shown in Figure 30 for the fourth load stage of case no. 1 where the source power factor drops below 0.97. However, if the source power factor stays above the pre-set power factor value of 0.97 during the load stage where the load power factor is greater than 0.97, then no action will be taken as shown in Figure 32 for the fourth load stage of case no. 3. The time required for the controller to improve the power factor in all three cases is relatively short compared with practical applications. In real cases, the power factor does remain unchanged for relatively longer time. The maximum time for power factor correction was 0.35 second recorded in test case no.1. Overall, the graphs show that the controller works satisfactory under different load conditions and when there is no need for capacitor compensation.

As mentioned before, power factor correction is really an important issue. The designed controller presented in this section shows an efficient, fast and accurate technique in reactive power control. As seen from the overall structure of the controller, it is applicable for lagging power factor loads. Practically, this is almost true but not always where at rare occasions the power factor of the total load is leading not lagging. This will bring the attention towards generalizing the presented controller such that it will work for both cases. Several issues are also need to be considered in the future such as the dynamics of motors. As known that most connected loads are motors which really necessitate testing this controller under these circumstances. From the test results, it was seen that the speed of the controller depends on the system time constant and hence, a time delay is needed to assure that the dynamics of the motors reach its equilibrium. Other issues related to Thyristors such as the harmonics are also need to be taken care by describing a harmonics filter. In addition to that, protection devices such as relays need to be checked during the controller action. Finally, the work presented was based on a single phase and it can be extended for three phase system.

3. Trending and prediction

Most of the more advanced prediction techniques can be subdivided into two separate tasks. In a first step, the modelling step, the algorithm uses a set of training data to identify a model of a process, from which the training data could have been obtained. In a second step, the simulation step, the algorithm uses the previously identified model to make predictions outside the training data set. The modelling algorithm can either attempt to identify the true structure of the system, from which the training data were obtained, or it can content itself with identifying any process able to explain the training data set. In the former case, we talk about a deep model, whereas models in the latter category are referred to as shallow models.

The identified model can be either a quantitative or a qualitative model. A quantitative model operates on the measurement data directly, whereas a qualitative model first discretizes the measurement data, and then reasons about the discrete classes only. Also the model can be either a parametric model or a nonparametric model. A parametric model maps the knowledge contained in the training data set onto a set of model parameters. During the simulation phase, the training data are no longer needed, since the information contained in them is now stored in the parameter values. A non-parametric model only classifies the training data during the modelling phase, and refers back to these classified training data during the simulation phase.

Fuzzy logic are used now a day in many application for diagnostic, prediction forecast and understanding the behaviour of very nonlinear systems such as marketing, electrical load forecast, work load analysis, technical analysis etc...

In this chapter Section 2.1, we shall introduce an important algorithm for classifying "clustering" the data based on fuzzy logic. Then two new fuzzy trending and prediction application shall be discussed. In Section 2.2 Accident rates Estimation Modelling Based on Human Factors shall be introduces, and in the next section 2.3, Fault Location in Distribution Networks shall be introduced.

3.1. Clustering algorithm and validity criteria

Clustering attempts to assess the relationships among patterns of the data set by organizing the patterns into groups or clusters such that patterns within a cluster are more similar to each other than are patterns belonging to different clusters. Many algorithms for hard and fuzzy clustering have been developed to accomplish this[42]. An intimately related important issue is the cluster validity, which deals with the significance of the structure imposed by a clustering method [43].

For fuzzy sets, the following definitions are recalled from [36]:

a. A fuzzy set in a universe discourse U is characterized by a membership function $\mu_A(x)$ that takes values in the interval [0,1].

b. Let $X=\{x_1,\ldots,x_n\}$ be any set, V_{cn} be the set of real $c\times n$ matrices $U=[\mu_{ij}]$, c, i, j be integer numbers with $2 \leq c \leq n$, $1 \leq i \leq c$ and $1 \leq j \leq n$ .Then the fuzzy partition matrix for X is the set $M_{fc} = \{U \in V_{cn} \mid \mu_{ij} \in [0,1]\}$ (32)

Such that

$$\sum_i \mu_{ij} = 1, \forall_j, 1 \leq j \leq n. \tag{33}$$

An α-cut of fuzzy set A is a crisp set $A\alpha$ that contains all the elements in U that have membership value in A greater thanα, that is

$$A_{\alpha} = \{x \in U \mid \mu_A(x) \geq \alpha\} \tag{34}$$

c.   Defuzzification is defined as a mapping from fuzzy set B` in V⊂ R to crisp point y* ∈ V. Conceptually, the task of the defuzzification is to specify a point in V that best represents the fuzzy set B`.

The following three criteria should be considered in choosing the defuzification method:

- Plausibility: The point y* should represent B` from an intuitive point of view.
- Computational simplicity.
- Continuity: A small Change in B` should not result in a large change in y*.

3.1.1. Fuzzy C-Means Clustering Algorithm(FCM)

The fuzzy c-means (FCM) clustering algorithm is the fuzzy equivalent of the nearest hard clustering algorithm [43,44], which minimizes the following objective function with respect to fuzzy membership μ_{ij}, and cluster centroid V_i.

$$J_m = \sum_I^c \sum_j^n (\mu_{ij})^f \mid\mid \left(X_j, V_i\right)\mid\mid^2 \tag{35}$$

where $X = [X_1,....,X_n]^t$ is a vector representing the data, c is the number of clusters, n is the number of data points and f is a fuzziness index (greater than 1)

The FCM algorithm is executed by the following steps:

a.   Initialize memberships μ_{ij} of X_j belonging to cluster i such that

$$\sum_j (\mu_{ij}) = 1 \tag{36}$$

b.   Compute the fuzzy centroid V_i from i=1 to i=c using

$$V_i = \frac{\sum_j (\mu_{ij})^m \times X_j}{\sum_j (\mu_{ij})^m} \tag{37}$$

c.   Update the fuzzy memberships μ_{ij} using

$$\mu_{ij} = \frac{(\mid\mid (X_j, V_i) \mid\mid)^{(-2/(m-1))}}{\sum_j (\mid\mid (X_j, V_i) \mid\mid)^{(-2/(m-1))}} \tag{38}$$

d.   Repeat steps 2 and 3 until the value of J_m is no longer decreasing.

The FCM always converges to strict local minimum of Jm starting from an initial guess of μ_{ij}, but different choices of initial μ_{ij} might lead to local minima.

3.1.2. Cluster validity

The quality of a clustering is indicated by how closely the data points are associated to the cluster centers, and it is the membership functions, which measure the level of association or classification. If the value of one of the membership is significantly larger than the others for a particular data point, then that point is identified as being a part of the subset of the data represented by the corresponding cluster center. But, each data point has c memberships; so, it is desirable to summarize the information contained in the memberships by a single number, which indicates how well the data point is classified by the clustering. This can be done in a variety of ways; for example, for the data point X_j with memberships $\{\mu_{ij}, ..., \mu_{cj}\}$, one could use any of the following:

$$Index1 = \sum_i (\mu_{ij})^2 \tag{39}$$

$$Index2 = \sum_i \mu_{ij} \log(\mu_{ij}) \tag{40}$$

$$Index3 = \max_i (\mu_{ij}) \tag{41}$$

$$Index4 = \min_i (\mu_{ij}) / \max_i (\mu_{ij}) \tag{42}$$

In fact, theses four indices of these are used as measure of the quality of clustering and are the basis for the *validity functional, partition coefficient, classification entropy, and proportion exponent*, respectively.

To illustrate the use of validity functional, we shall focus on the partition coefficient technique because of its simplicity. It is based on using $S_j = \sum_i (\mu_{ij})^2$ as a measure of how well the jth data point has been classified. This is a reasonable indicator because the closer a data point is to a cluster center, the closer S_j is to 1, the maximum value it could have. Conversely, the further away the kth point is from all the cluster centers the closer the value of S_j is to 1/c, the minimum possible value. The partition coefficient is then the average over the data set of the S_j's. In particular, for a data set $X=\{x_1,...,x_i\}$ and a specific choice of c and m one obtains the output of fuzzy c-means and computes the partition coefficient (PC) by $PC=\sum_j(\sum_i(\mu_{ij})^2/n$. The closer this value is to one the better the data are classified. So, in theory, one computes PC for the outputs of a variety of values of c and m selects the best clustering as the one corresponding to the highest partition coefficient [44,45].

3.2. Accident rates estimation modeling based on human factors using fuzzy c-mean clustering techniques

Several individual books [37] and projects shed light on worker accident causation. One study on the Bonneville Dam project, reported that seven times the number of work accidents that had occurred on this project were due to unsafe employee actions rather than to unsafe site conditions. In addition, this study found that the negative attitude of the workers toward safety was a major factor in accident occurrence.

In [38]. Many organizations spend a lot of time and effort trying to improve safety. As well as addressing technical and hardware issues, many conduct safety management system audits to discover deviations from the performance standards set in their Health & Safety

Policies. Line management is encouraged to conduct regular inspections of the workplace and employees are trained to behave safely and are given the appropriate protective equipment. The impact of such initiatives could be seen in the overall downward trend in accident statistics from 1990 to 1998/99. After 2000, accident statistics started rising in many UK industrial sectors. In the Quarry Industry, for example, there has been a 60% rise in the number of fatalities.

Another study by Stanford University [39] indicated that risk taking is often a normal part of human psychology. We sometimes drive too fast or take chances we should not. Risk which is taken on the job site, however, can be fatal. This study found that many workers believe that taking unnecessary risks is an accepted part of the job process. This risk acceptance attitude leads to carelessness and accidents. The results of the Stanford study show that workers who are likely to have lost-time accidents share similar characteristics. These workers have a negative attitude toward doing their jobs safely, and they accept unnecessary risk and, therefore, do not work safely. Taking unnecessary risks and adopting a poor safety attitude simply makes workers more prone to accident occurrences. The conclusion of the Stanford study clearly supports the contention that employee actions and attitudes can affect the number and type of workplace accidents. Employers can and do address this attitude of risk taking through safety education, safety rules, and training programs. However, no employer can supervise each employee every minute of the work day.

In [40], the paper focuses on the development and representation of linguistic variables to model risk levels subjectively. These variables are then quantified using fuzzy set theory. In this paper the development of two safety evaluation frameworks, using fuzzy logic approaches for maritime engineering safety based decision support in the concept design stage are presented. An example is used to illustrate and compare the proposed approaches. The paper also suggests that future risk analysis in maritime engineering applications may take full advantages of fuzzy logic approaches to complement existing ones.

The field of fuzzy systems has been making rapid progress over the past decade [36]. There are two kinds of justification for fuzzy systems to be used to achieve our objective:

a. The problem is too complicated for precise description to be obtained, therefore approximation, or fuzziness, must be introduced in order to be a reasonable and net traceable model.
b. As we move into the information era, knowledge is becoming increasingly important and the need for a theory to formulate human knowledge in a systematic manner becomes the norm not the exception.

But as a general principle, a good engineering theory should be capable of making use of all the available information effectively. For many practical systems, important information comes from two sources: one source is from human experts who describe their knowledge about the system in natural languages; the other is sensory measurements or mathematical models that are derived from to physical laws.

An important task, therefore, is to combine these two types of information into system designs. Therefore, the key question is how to transform human knowledge base into a mathematical formula or model. Essentially, what a fuzzy system does is to perform this transformation in a systematic way.

In this part of Section 2, we attempt to use a completely different approach to analyze - accidents. A model shall be developed for data collected from an accident rate questionnaire filled-in by laborers working for a reputable construction company. This questionnaire was designed to include information about human factors, as well as other factors such as work type, managerial factors, training, physical factors and the historical accident rate for each labor during his period of employment in this particular construction company, and his experience during his career life time. The collected data shall be split into a training set for model construction and a test for model verification. The training information shall be classified into a number of groups or clusters, the centroids of these clusters were subsequently used to generate a set of rules to develop a fuzzy engine, which can then predict and forecast the rate of accidents. The test cases shall be used to verify and validate the developed model. Discussion on the results obtained from using fuzzy logic techniques shall be carried out.

3.2.1. Data organization

Construction sites are very dynamic and complex by nature, creating the potential for hazards that change constantly. So, what was safe yesterday may no longer be safe today. Thus, safety precautions should be followed and controlled. Unsafe working conditions and accidents are usually warning signs that something is wrong and has to be rectified.

Different government authorities measure safety at construction sites [41], however co-ordination and sharing of information with each other is still lacking. In addition, the data available on construction site accidents are neither accurate nor complete, due to the absence of a reliable accident reporting and recording system. Incomplete records are due to the poor accident investigation that may be a result of:

- Reluctance of reporters to assert authority.
- Inexperienced and untrained investigators
- Narrow interpretation
- Judgmental behavior.
- Incomplete or erroneous conclusions.
- Delays in accident investigations.

For these reasons we endeavored to avoid the normal way of doing the job, and instead, we focused on the laborer, himself, and his accident rates during his years of work experience as an expert source of data. We have tried to design the questionnaire in such a way as to serve our purpose of analyzing the data, and selected the interview method to get the maximum precise data possible.

3.2.1.1. Questionnaire design

In the design of questionnaire (Appendix) we have selected some certain features of human nature that, we believe, have a great potential on the accident causation in the local market [41]. The first page of the questionnaire concentrated on the personal information of the workers: i.e. 'height', 'weight', 'optical status', 'hearing ability', 'general health', 'education' and 'adherence to safety rules'. In this part we used some linguistic evaluations like 'high', 'low', 'fair', 'good', 'medium' etc., and in some others we have used numerical evaluations like in height, weight, as well as education. Since the objective here is to create a fuzzy model, high accuracy is not important and we considered that the respondents from the same field made the same judgments.

The second page was designed to concentrate on the work information: 'overtime work', 'experience', 'work nature', 'work type', 'hazardous level', 'needs for safety-gears', 'work location' and 'level of boredom'. Again, we have used some linguistic evaluations as well as numerical evaluations. We also concentrated on the managerial factors: 'salary received on time', 'level of training' 'importance' and level of importance placed on safety', with only linguistic evaluations.

The third page was a mix of both external factors: 'noise', 'live with family' and 'communication 'language', and accident history focusing on the number of accidents the laborer has faced during his work in the local construction market, which was the most important data that we needed to develop our model. The severity of accidents has not been taken into account since it is not considered a factor influencing the accident rate.

The cases obtained for this study were collected from three different construction companies selected to represent the local market. We have tried to select the cases from different ranks of the workforce, from higher levels to the lower levels to be able to study the different accident level cases that serve the purpose as well as adding versatility and diversity to this work.

3.2.1.2. The response

From the original cases that we collected on 95 people, we included only 76 cases and excluded 19 cases, which were incomplete. This has produced a very high response rate that reached 82.1%, which is relatively high, especially as the questionnaire is lengthy and a little bit complicated. More cases could have been obtained. However, since the aim of this study was to develop a model for accidents, the number of cases is not set a priority. Rather, the cases are accumulated and the algorithms are stopped when a cluster validity criterion is satisfied thereby yielding the optimum number of clusters.

3.2.1.3. Limitations

Limitations in this study should be noted. One of the limitations is that we did not include any specific information related to the accident consequences. Another limitation is that this study was made only on males and no female cases were studied, which makes it specific only to one sex.

3.2.1.4. *The feature matrix*

The feature matrix (FM) is the most important part of our work in this section , since by using this matrix we have been able to convert the linguistic variables into numerical variables. Thus, we can deal easily with practical cases and reduce the required operations of processing the output.

The columns of the FM matrix represent the feature variables which we obtained from the questionnaire. The rows of the matrix represent the different cases of laborers that we selected for interviews. Thus, for each case in the matrix we mapped the linguistic meanings into numbers according to weights we have proposed. For example, in case labelled (S1), the feature weight 1 (FW1) represents the rate of accidents per year of experience, which is the actual representation of the accident rates. FW2 represents the ratio between weight and height (specific weight). FW3 represents optical status and FW4 represents hearing ability. These feature weights are scaled on a scale of five from 1 to 5, to represent the linguistic variables. Therefore, 1 means `very bad`, 2 means `bad`, 3 means `medium`, 4 means `good` and 5 means `very good`. All the other variables are dealt in the same way until the matrix was generated. A sample of the feature matrix is shown in Table3.

Feature Case No	Accidents/ experience	Weight/ Height	Optical status	Hearing Ability	General Health	Adherence to Safety	Education	Overtime work
S1	1/12	71/160	6/18	5	5	4	5	2
S2	5/12	77/ 170	6/60	5	4	3	5	3
S3	1/21	90/175	6/6	4	5	5	5	5
S4	0/8	54/165	6/60	4	5	2	5	20
S5	6/3.5	68/187	6/36	3	4	3	5	0
S6	10/11	85/177	6/6	4	4	3	5	10
S7	2/19	76/173	6/60	4	5	5	5	14
S8	18/25	72/170	6/18	3	4	4	4	4
S9	45/14	80/176	6/6	4	4	3	4	8
S10	3/20	81/174	6/6	4	4	5	4	1

Table 3. Sample of feature matrix illustrates the weight of some features for the first 10 cases

3.2.2. Modelling

In this stage, FCM techniques will be implemented on the feature matrix after normalizing the data, based on column maximum values for ease and as being more indicative, then deciding the optimum number of clusters, by applying cluster validity techniques. The centroides for these optimum clusters are considered perfect models represent the feature matrix.

In order to obtain the models, the following steps have been implemented:

- Each column in the feature matrix is normalized by dividing all the numbers in this column by the maximum number of the absolute values of all the numbers in the said column.
- Cluster validity study is implemented to determine the optimum number of clusters for the normalized data.
- FCM technique is implemented to determine the centroids matrix of the selected number of clusters.

MATLAB fuzzy toolbox has been used to implement the above three steps [46].

3.2.2.1. Clustering results and discussions

The results can be summarized as follows:

a. The optimum number of cluster (twelve) is determined by implementing cluster validity technique. The result is illustrated in Figure 33 which gives the relation between the number of clusters and the corresponding error where:

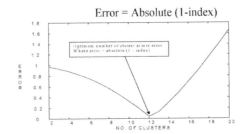

Figure 33. Relation between number of clusters and the corresponding error

b. Table 9 (Appendix) illustrates the centroide matrix for the optimum number of clusters (twelve), where FW(j) stands for feature weight described in Table 10 (Appendix).

3.2.2.2. Scaling of data

In order to reduce the error in the estimation of the membership functions ranges (e.g. the estimated range of FW5 was from 1 to 5, while from centroide matrix the range for the same feature is found to be from 3.750 to 4.9953) each feature vector element is scaled according to the following formula:

$$X_{scaled} = X_i - X_{min} / (X_{max} - X_{min})$$ (43)

Where:

X_i = Vector element
X_{min} = Min. Vector element
X_{max} = Max. Vector element

3.2.2.3. Model development

By comparing each row of the centroids matrix with the contents of the questionnaire after scaling, one can infer the structure of the respective model, for example maximum and minimum rate of accident as illustrated in Table 11 (Appendix). From structure, following features can be extracted:

a.   Labours in this construction company can be classified into 12 models.
b.   The expected range of accident rate in company varies from 0.1581 to 2.8894 accidents per year.
c.   By comparing the above two extreme models, one can easily discover that receiving salary on time has no effective impact on accident rate.
d.   The highest rate of accidents occurs to non-local workers (Live without their family most of the time).
e.   The twelve models obtained above shall be utilized later as fuzzy rules representing the this local construction company to predict the rate of accident for any person working on construction field.

3.2.3. Accident rate prediction

Now, fuzzy logic techniques will be implemented using the models obtained in Section 2.2.2, as perfect fuzzy rules, to predict the accident rate for any laborer who works in the construction field. The following flow chart (Figure 34) describes the fuzzy accident prediction system:

The models obtained from the fuzzy c -means the clustering process has been considered as very good and suitable fuzzy rules that govern the relation between the laborers in construction field and the expected annual rate of accident.

The beauty of using this type of clustering is not only to achieve the required models, but also these models are fuzzy, and can be geared in the fuzzy engine.

In order to fuzzify the model variables, a suitable number of Gauss functions (linguistic variable) is selected for each linguistic value so that any rule must fire all the linguistic values and the rules are given as follows [36]:

$$\text{IF } \left(FW1 \text{ is } mf_a^{\,1}\right) \text{ and } \left(FW2 \text{ is } mf_b^{\,2}\right) \text{ and} \ldots \left(FW22 \text{ is } mf_v^{\,22}\right) \text{ THEN } \left(FW23 \text{ is } mf_w^{\,23}\right) \quad (44)$$

Where,

FW(1to22) : input linguistic variable
FW23 : output linguistic variable
mf $_{a,b\,..w}$: semantic rule
a,b ..w: integer number from 1 to 5

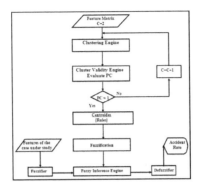

Figure 34. Fuzzy Accident Prediction Flowchart

Mamdani inference engine with centroid defuzzification and proportional aggregation is used in the construction of the fuzzy system. MATLAB Fuzzy Toolbox has been used to implement the required fuzzy prediction system. Eight cases have been tested and the results are given in hereinafter.

3.2.3.1. Relevant results

a. Eight additional random cases have been chosen from a local construction company. Features, actual accident rate and output of fuzzy prediction system for each test case are given in Table 12 (Appendix). The standard deviation between the actual accident rate and the predicted accident rate for each test case is calculated then the average standard deviation is obtained in order to determine the validity of the model. The results are shown as follows in Table 4:

Case Number	Standard Deviation
Case#1	0.3536
Case#2	0.2828
Case#3	0.4243
Case#4	0.7071
Case#5	01414
Case#6	0.1414
Case#7	0.2121
Case#8	0.0318
Average	0.2868

Table 4. Standard deviation results for the test cases

b. In (Appendix), Figure 37 shows the output of the prediction system for a case that fires all the linguistic values at the middle. It is observed from this result that the laborers with `average` personal information, `average` work condition, `average` managerial condition and `average` external effects are exposed to `average` annual accident rate.

c. Figure 38(Appendix) correlates the laborers general health and specific weight with their annual accident rate considering all other factors are `average`. It is clear that the annual accident rate increases with the significant increase of the specific weight and significant poorness of the general health.

d. In Figure 39 (Appendix), the laborers educational level and safety adherence are plotted against their annual accident rate considering all other factors are `average`. It is noticed from this illustration that the laborers with the two educational level extremes are exposed to accidents more than the `average` educational level laborers.

e. Figure 40 (Appendix) correlates the laborers optical status and hearing ability with their annual accident rate considering all other factors to be `average`. It is clear that the annual accident rate increases with the significant poorness of the optical status and hearing ability.

f. From the last plot (Figure 41 - Appendix), it is clear that if the company is not keen on the level of safety, the annual accident rate will be increased, considering all other factors are `average`. In addition, it is noticeable from the figure that the tasks that are considered to be dangerous and need safety-gear are a source of accidents.

3.2.4. Discussion

The main objective of in this example to generate a human model which reflects the interaction between human factors in addition to other factors such as managerial factors, accident information, and work information, using fuzzy clustering techniques. Secondly, to predict annual rate of accident for any sample of workers by applying fuzzy logic techniques. Some of the important results obtained can be summarized as follows:

a. Fuzzy clustering techniques can be used to build a model that characterizes the different features of workers in local construction field against their rate of accidents.

b. The model obtained from clustering is considered as rules to be used in a fuzzy logic engine to predict the rate of accidents for any worker in the construction field.

c. For any specific case, 231 correlations (between any two features and the rate of accidents) can be done via the fuzzy engine.

d. Optimum training to improve the safety attitude for certain laborer with minimum cost can be estimated by analyzing the correlation between the level of training and rate of accident of this particular labourer.

e. By analyzing the correlation between level of safety importance and rate of accidents for the workers in a particular company, the limits of the effective safety improvement can be predicted in order to evaluate the investment in this direction.

f. A similar technique can be applied to a particular company to predict the rate of accident in order to estimate the insurance rate for the people who work in this particular company.

g. Using the accident rate fuzzy prediction techniques companies can select the most suitable workers for any particular task

3.3. Fault location in distribution networks using fuzzy c-mean clustering techniques

In last section of this chapter we shall studies an existing 13.8 kilovolt distribution network which, serves an oil production field spread over an area of approximately sixty kilometers square, in order to locate any fault that may occur anywhere in the network using fuzzy c-mean classification techniques.

In addition, we shall introduce several methods for normalizing data and selecting the optimum number of clusters in order to classify data. Results and conclusion shall be also given to show the feasibility for the using the fuzzy logic to locate the fault location.

3.3.1. Network description

A joint venture oil company possesses two production areas, Area1 and Area2. For each area a power distribution system is provided. The power supply required for the two fields is provided from 130 MVA capacity power generation plant located in Area1. Twenty MW is transmitted to Area2 via 35KM long over head transmission line (OHTL) on wooden poles. Area2 power system also contains two 3.16 MVA stand by generators.

Area2 distribution system consists of three radial overhead transmission lines, 8-10 KM long each, serve submersible oil pumps and other loads scattered in the field (60-kilometer square). These overhead transmission lines have neither differential relays nor sectionalizing fuses. Programmable logic Controller (PLC) is connected to the incomers, outgoings and generators auxiliaries at Area2 power station. This PLC records the running and tripping information for all bus-bar compartments.

Due to the aging of the system, remote area problems and harsh desert weather, repeated faults are experienced in the grid. Because of unavailability of differential relays and/or sectionalizing fuses, it is very difficult and long time is consumed to locate any fault in this network. The problem is reflecting passively on the oil productivity of this important area.

3.3.2. Feature Matrix

In order to measure the features of the faults at 176 nodes of the network; load flow study is implemented to determine the respective power loss for each short circuit case and also short circuit study is carried out to determine the feature vector for each short circuit case.

The results, obtained from the load flow study and the short circuit study, shall be used to form the network *feature matrix,* which will be clustered and analyzed hereinafter. The parameters that selected to build the feature matrix are shown in Table 5.

Fuzzy clustering technique shall be used to classify the possible fault locations, which can be near to any node in the network or near to a chosen set of nodes based on the operator experience, into groups. The optimum number of groups (clusters) is computed using validity clustering technique. The remaining 13 cases are used as test cases. Euclidean distance technique is implemented to find out the group of nodes, which the fault may be found near to, for each test case.

Feeder 1	Feeder 2	Feeder 3
Set of nods fed from Feeder 1	Set of nods fed from Feeder 2	Set of nods fed from Feeder 3
Circuit breaker 1 status	Circuit breaker 2 status	Circuit breaker 3 status
Feeder 1 Short circuit Current red from substation	Feeder 2 Short circuit Current red from substation	Feeder 2 Short circuit Current red from substation
Phase Angel A1	Phase Angel A2	Phase Angel A3
Phase Angel B1	Phase Angel B2	Phase Angel B3
Phase Angel C1	Phase Angel C2	Phase Angel C3
Power dip in Feeder 1	Power dip in Feeder 2	Power dip in Feeder 3
VAR dip in Feeder 1	VAR dip in Feeder 2	VAR dip in Feeder 3

Table 5. Summary of the parameters that are selected to build the feature matrix.

Assumptions: The following assumptions are considered:

a.   The temperature of the network conductors is assumed constant at seventy degree Celsius.
b.   Only symmetrical short circuit is conceded. The same procedures can be implemented for any other type of faults.

Assumption (a) is valid since the transmission lines are short [47]. For assumption (b), the same work can be repeated for all other types of failures.

3.3.3. Fault location using column maximum normalization

Now, FCM technique can be implemented after normalizing the data based on column maximum values and deciding the optimum number of clusters. The results shall be analyzed in order to locate any failures may occur in the network.

3.3.3.1. Calculation procedures

To detect the fault using FCM technique with column maximum normalization the following steps have been implemented:

a.   Each column in the Feature Matrix is normalized by dividing all the numbers in this column by the maximum number of the absolute values of all the numbers in the said column.
b.   Cluster validity study is implemented to determine the optimum number of clusters for the normalized data.
c.   FCM technique is implemented to determine the fuzzy partition matrix of the selected number of clusters.

d. The norms between each test data and the full data in the chosen cluster is examined, accordingly the corresponding cluster for each test data is decided based on the minimum norm obtained from the said examination.
e. Alpha-cut defuzzification is used with alpha equal to 90% of the average of norms between the cluster center and its data.
f. The nearest node to the fault is checked in order to determine whether it is included in the possible locations or not.

3.3.3.2. Results and discussion

Matlab program is written to implement the above six steps and the results are analyzed and summarized as follows:

a. The optimum number of cluster (twenty five) is determined by implementing cluster validity technique discussed in 2.1. The result is illustrated in Fig 35 which give the relation between the number of clusters and the corresponding error where:

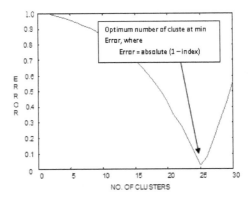

Figure 35. The relation between number of clusters and the corresponding error

b. Table 6 shows the effort saved to locate 13 fault cases. From this table, it can be noticed that:
   - Effort Saving = 1-Ratio between the number of possible locations to the total number of nodes*100%. (45)

Also, we can define the following terms:

   - Percentage of successful trials (Ratio of the cases of including the nearest node in the possible locations to the total number of testing cases*100%) = 92%. (46)
   - Average effort saving (Summation of the Effort Saving percentages divided by the total number of the cases) =75% (47)

It is important to notice that fuzzy cluster technique failed to locate the fault of test data number 4 due to the lack of information near to this location. However, it is expected that for more available information the performance of this technique will be improved.

TESTING CASE NO.	NUMBER OF POSSIBLE LOCATIONS	NEAREST NODE EXISTS IN THE POSSIBLE LOCATIONS	EFFORT SAVING
1	38	Yes	77%
2	38	Yes	77%
3	23	Yes	86%
4	2	No	0%
5	35	Yes	79%
6	56	Yes	66%
7	23	Yes	86%
8	49	Yes	70%
9	23	Yes	86%
10	17	Yes	90%
11	20	Yes	88%
12	20	Yes	88%
13	17	Yes	90%

Table 6. Effort Saving

3.3.4. Fault location using simple maximum normalization

Here, FCM technique is implemented after normalizing the data based on the maximum value of the data and deciding the optimum number of clusters. Then, the results are analyzed in order to locate any failures may occur in the network.

3.3.4.1. Calculation procedures

To detect the fault using FCM technique with simple maximum normalization, the following steps have been implemented:

a. All the numbers in the feature matrix are normalized by dividing all of them by the maximum number of the absolute values of all the numbers in the matrix.
b. The data are preliminary classified into clusters based on the understanding of the network operation.
c. Cluster validity study is implemented to determine the optimum number of clusters for the selected cluster.
d. FCM technique is implemented to determine the fuzzy partition matrix for the selected number of clusters.
e. The norms between each test data and the full data in the chosen cluster is examined, accordingly the corresponding cluster for each test data is decided based on the minimum norm obtained from the said examination.
f. Alpha-cut defuzzification is used with alpha equal to the average of norms between the cluster center and its data.
g. The nearest node to the fault is checked in order to determine whether it is included in the possible locations or not.

3.3.4.2. Results

Matlab program is written to implement the above six steps and the results are analyzed and summarized as follows:

a. The optimum number of cluster is determined by implementing cluster validity technique and the results are given in Table 7. It is clear from the Table that the optimum number of clusters varies from case to another, which indicates that for any considerable additional of information, cluster validity study should be implemented again to find the new optimum number of clusters.

CASE NUMBER	CASE DISCRIPTION	OPTIMUM NUMBER OF CLUSTERS
1	POWER DIP IN FEEDER #1 AND C.B.1 TRIPS	13
2	POWER DIP IN FEEDER #2 AND C.B.2 TRIPS	15
3	POWER DIP IN FEEDER #3 AND C.B.3 TRIPS	7
4	POWER DIP IN FEEDER #1 AND C.B.1 DOES NOT TRIP	10
5	POWER DIP IN FEEDER #2 AND C.B.2 DOES NOT TRIP	11
6	POWER DIP IN FEEDER #2 AND C.B.2 DOES NOT TRIP	5

Table 7. Shows the optimum number of clusters for each case identified by the operator

b. Figure 36 illustrates the relation between the number of clusters and the corresponding error in case 1, where error is calculated as given in (45).

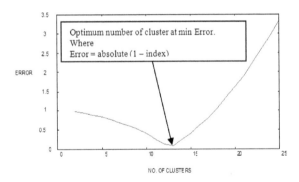

Figure 36. Shows the relation between number of clusters and the corresponding error for case 1

c. Table 8 shows the effort saved to locate 13 fault cases. From the this table , it can be noticed that:
  • Effort Saving can be calculated as given in (45)
  • Percentage of successful trails = 100%. See (46)
  • Average effort saving = 87% See (47)

TESTING CASE NO	NUMBER OF POSSIBLE LOCATIONS	NEAREST NODE EXISTS IN THE POSSIBLE LOCATIONS	EFFORT SAVING
1	21	Yes	87%
2	16	Yes	90%
3	23	Yes	86%
4	21	Yes	87%
5	34	Yes	79%
6	14	Yes	91%
7	33	Yes	80%
8	30	Yes	82%
9	33	Yes	80%
10	5	Yes	97%
11	10	Yes	94%
12	10	Yes	94%
13	9	Yes	94%

Table 8. Effort saved

3.3.5. Discussion

The main objective of was to apply fuzzy c-mean clustering technique to locate any 3-phase fault that may occur at any point on actual power distribution network. Two different normalizing methods have been used to process the feature matrix data. The result obtained can be summarized as follows:

a. Fuzzy clustering technique can be used to investigate the location of faults in networks.
b. Any actual fault can be utilized and be fed back to the database of the clustering system to improve its performance and efficiency.
c. Understanding the network configuration and operation can be utilized to improve the clustering which gives better results.
d. Data matrix comprising the feature vectors should reflect good and adequate description of the network.
e. For any major network's upgrading or change the clustering should be implemented again by using the new data obtained from complete study of the modified network.

4. Summary of the chapter and conclusion

In this chapter we presented five new problems from different application; control, accident analysis, process and electrical network. By using fuzzy logic technique, we succeeded to resolve these problems efficiently. We also introduced different type of normalization of the data. Generating fuzzy rules by either linearization of the curves or by clustering the data were presented as well. Then a method of coloration and prediction of information using the generated fuzzy rules were provided.

Appendix

Appendix 1

Questionnaire

Case No.:

1.Personal Information

Age : 20-30 30-40 40-50 20-60

Sex : Male Female

Social Status: Bachelor Married

Height: (cm)

Weight: (kg)

Optic Status: 6:6 6:18 6:36 6:60

Hearing Ability: Very bad Bad Medium Good Very good

General Health: Very bad Bad Medium Good Very good

Adherence to safety rules: Very high High Fair Low Very low

Education: Illiterate Year 1-4 Year 4-8 Year 8-12 University

2.Work Information:

Overtime Work Hours per week

Experience: Years

Work Nature: Mental% Manual%

Work type: Very Easy Easy Fair Tough Very Tough

Hazardous level: Very low Low Medium Hazard Very hazard

Needs of safety gear: No need Rear Sometime Always

Work location: Indoor% Office work% Outdoor%

Level of boredom: Very high High Fair Low Very low

3. Managerial Factors:

Salary received on time: Strongly disagree Disagree Agree Strongly agree

Level of training importance: Very low Low Medium High Very High

Level of safety importance: Very low Low Medium High Very High

4.External Factors:

Noise level: Very low Low Medium High Very high

Live with family Months per year

Communication Language: Very bad Bad Medium Good Very good

5.Accident Information:

Accident History: (How many accidents did the case have?)

FW1	FW2	FW3	FW4	FW5	FW6	FW7	FW8	FW9	FW10	FW11	FW12
FW13	FW14	FW15	FW16	FW17	FW18	FW19	FW20	FW21	FW22	FW23	
0.4899	0.4824	0.5548	4.3318	4.3324	3.9963	4.3311	15.6376	46.6328	53.3672	3.3330	3.6684
3.6677	8.3954	19.9348	71.6699	2.6679	2.6683	3.0055	4.0012	3.3330	5.0219	3.0031	
0.3425	0.4881	0.3333	4.9983	4.3328	4.3310	4.6673	0.0000	94.9920	5.0080	3.9983	3.9999
3.3345	23.3133	6.7051	69.9817	2.9999	3.3311	3.3364	4.0020	3.0000	12.0000	3.6656	
0.1581	0.5055	0.7761	4.5545	4.6684	3.9956	4.7782	4.2714	96.6805	3.3195	3.5547	3.7724
3.1099	7.8244	45.6152	46.5605	2.8862	2.3357	3.4470	3.9988	3.6650	11.3039	4.1094	
2.8894	0.4899	0.5016	4.9953	4.9953	2.5048	3.0046	7.4857	2.5093	97.4907	3.9965	4.4954
2.9979	5.0231	14.9297	80.0472	4.4941	2.5000	2.0000	1.5024	3.0034	6.5188	3.0001	
0.3674	0.4435	0.7917	4.7499	4.4997	3.9994	4.2500	8.7491	47.4970	52.5030	3.7497	3.7497
2.7503	21.2531	28.7504	49.9965	3.2497	3.0000	2.7496	3.9992	3.7492	3.3779	3.5003	
0.6164	0.4388	0.8158	4.3160	4.3687	2.7888	3.3152	8.4245	5.0013	94.9987	3.8953	3.8950
3.2638	2.3621	0.0000	97.6379	2.9997	2.7371	2.2635	2.8957	3.4735	4.2838	3.2107	
1.2917	0.4321	0.5000	4.0000	3.7500	3.0000	3.7500	7.5000	27.5000	72.5000	4.0000	3.2500
3.0000	71.2500	10.0000	18.7500	3.2500	2.5000	2.5000	2.5000	3.7500	3.6250	3.0000	
0.3401	0.4600	0.4002	4.3754	4.3744	4.1238	4.8749	3.2411	71.2455	28.7545	3.4994	3.6244
3.2492	6.2440	61.2585	32.4975	2.4996	3.7507	3.0011	3.9998	3.1258	9.3725	4.6244	
0.8238	0.4696	0.2803	4.3322	4.3322	2.3357	4.0000	10.6568	6.7131	93.2869	3.6678	3.6678
2.6680	41.5930	11.7266	46.6804	2.6645	2.6676	2.0000	2.6649	3.6645	4.5258	3.6678	
0.5278	0.4539	0.5997	4.3992	4.2006	3.0011	4.6002	2.8065	62.0051	37.9949	3.0005	3.0005
3.2006	32.9938	43.0014	24.0048	3.4002	3.5998	2.5999	2.6004	2.4002	10.4017	3.8004	
0.2282	0.4517	0.3471	4.7537	4.8743	3.6330	4.8743	3.0413	97.4866	2.5134	3.3767	2.3770
2.3720	4.2728	83.2449	12.4822	2.8797	3.3770	4.1206	3.6233	2.4976	9.4162	4.5074	
1.1461	0.4654	0.8339	4.0001	4.2512	3.2512	3.2468	7.4926	27.4917	72.5083	3.8764	3.7486
3.3742	18.1000	13.1229	68.7771	2.8766	2.6225	3.5013	3.7523	3.5017	5.2219	3.2467	

Table 9. Illustrates the centroide matrix for the optimum number of clusters (twelve),

Feature Weight	Description	Feature Weight	Description
FW1	Accidents/ experience	FW13	Need for Safety Gear
FW2	Weight/Height	FW14	Indoor Work
FW3	Optical status	FW15	Office Work
FW4	Hearing Ability	FW16	Outdoor Work
FW5	General Health	FW17	Level of Boredom
FW6	Adherence to Safety	FW18	Salary on time
FW7	Education	FW19	Level of Training
FW8	Overtime work	FW20	Level of Safety
FW9	Mental work	FW21	Noise level
FW10	Manual work	FW22	live with family
FW11	Work type	FW23	Communication language
FW12	Hazard level		

Table 10. DESCRIPTION of feature weights,

	Personal Factors						Work Factors										Managerial Factors			External Factors		
	Specific weight	Optical status	Hearing ability	General health	Adherence to safety	Education	Overtime work rate	Mental work rate	Manual work rate	Work type	Hazardous level	Needs for safety-gear	Indoor work rate	Office work rate	Outdoor work rate	Level of boredom	Delay on receiving the salary on time	Level of training	Level of safety importance	Level of noise	Live with family	Communication language level
Minimum accident rate 0.1581	high	average high	average	above average	above average	high	low	high	low	fair	above average	average high	low	average	average	low	high	Very high	High	high	9-12 months per year	above average
Maximum accident rate 2.8894.	above average	low	high	good	low	low	average	low	high	tough	high	average low	low	under average	over average	high	high	low	Low	average low	4-6 months per year.	poor

Table 11. Features of the model of Maximum and Minimum accident rate

Feature	Case#1	Case#2	Case#3	Case#4	Case#5	Case#6	Case#7	Case#8
Accidents/ Year experience	2.6	3.2	2.5	2.3	3.1	0	1.1	0.125
Accident rate predicted	2.1	2.8	1.9	1.3	2.9	0.2	1.4	.17
Weight/Height	0.465	.454	.42	.466	.429	.49	.51	.415
Optical status	6/36	6/6	6/6	6/6	6/12	6/60	6/6	6/6
Hearing Ability	Medium	Good	V.Good	Good	Medium	V.Good	V.Good	V.Good
General Health	Good	Good	Good	Good	Good	V.Good	Good	V.Good
Adherence to Safety	Fair	Fair	Fair	V. High	Low	V. High	Fair	V. High
Education	16	9	12	8	5	17	10	19
Overtime work	0	8	2	1	8	0	3	0
Mental work	60	20	50	70	10	100	50	90
Manual work	40	80	50	30	90	0	50	10
Work type	Fait	Tough	Fair	Fair	Fair	V. Easy	Fair	V. Easy
Hazard level	V. High	High	Medium	V. High	High	V. Low	Medium	Low
Need for Safety Gear	Sometime	Sometime	Sometime	Always	Sometime	Rare	Rare	No need
Indoor Work	0	20	30	5	0	0	80	0
Office Work	80	30	40	45	0	100	20	100
Outdoor Work	30	50	30	50	100	0	0	0
Level of Boredom	Medium	High	Medium	Medium	High	Low	High	Medium
Salary on time	Strongly Agree	Agree	Agree	Strongly Agree	Disagree	Strongly Agree	Agree	Strongly Agree
Level of Training importance	Medium	Low	Low	Low	Low	V. High	High	High
Level of Safety Importance	High	Medium	Medium	High	Medium	High	High	V. High
Noise level	High	High	Medium	V. High	Medium	Low	Medium	Low
Live with family	12	12	6	12	0	9	12	12
Communication language	Good	V.Good	Medium	Good	Medium	V. Good	Good	V.Good

Table 12. Result of eight test cases

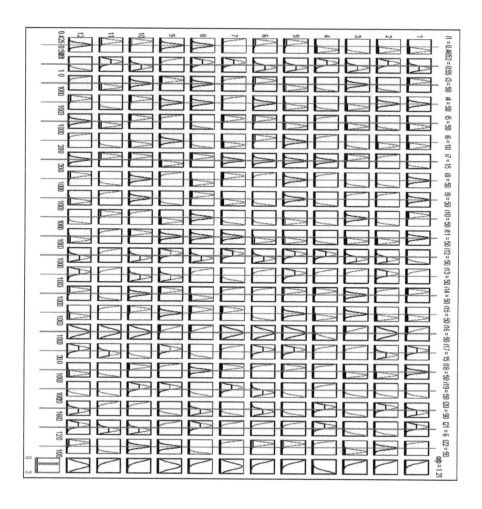

Figure 37. The output of the prediction system for `average` inputs.

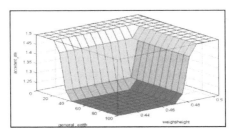

Figure 38. Correlation between weight/height, general health and rate of accident

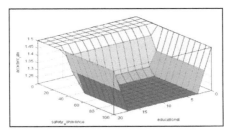

Figure 39. Correlation between education level, safety adherence and rate of accident

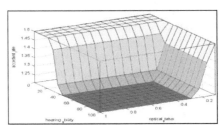

Figure 40. Correlation between optical status, hearing ability and rate of accidents.

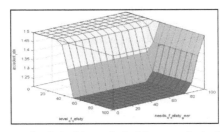

Figure 41. Correlation between level of safety, need of safety-gear and rate of accident.

Author details

Muhammad M.A.S. Mahmoud

Received the B.S. degree in Electrical Engineering from Cairo University and the M.Sc. degree from Kuwait University. PH.D Transilvania University of Brasov, Romania He occupies a position of Senior Engineer at Al Hosn Gas Co. UAE

5. References

[1] Zadeh, L.A., "Fuzzy Sets", Information and Control, Vol. 12, pp. 338-353, 1965.

[2] M. S. Mahmoud, "Computer-Operated Systems Control", Marcel Dekker Inc., New York, 1991.

[3] C. C. Lee, Fuzzy Logic in Control Systems: Fuzzy Logic Controller, Parts I& II", IEEE Systems, Man and Cybernetics, Vol. 20, No.2, March/April 1990, pp.404-435.

[4] M. S. Mahmoud, S.Z. Eid and A .A. Abou-Elseoud, "A Real Time Expert System for Dynamical Processes", IEEE Transactions Systems, Man and Cybernetics, Vol. SMC-19, No. 5, September/October 1989, pp. 1101-1105.

[5] M. S. Mahmoud, S. Kotob, and A. A. Abou-Elseoud, "A Learning Rule- Based Control System", Journal of Information and Decision Technologies, Vol. 18, No. 1, January 1992, pp. 55-66.

[6] M. S. Mahmoud, A. A. Abou-Elseoud and S. Kotob, "Development of Expert Control Systems: A Pattern Classification and Recognition Approach", Journal of Intelligent and Robotic Systems, Vol. 5, No. 2, April 1992, pp.129-146.

[7] L. X. Wang, "A Supervisory Controller for Fuzzy Control Systems that Guarantees Stability", IEEE Trans. Automatic Control, Vol. 39, No. 9, September 1994, pp. 1845-1847.

[8] K. M. Passino, and S. Yurkovick, "Fuzzy Control", Addison Wesley, California, 1998.

[9] Wang, Li-Xin, "A Course In Fuzzy Systems And Control", Prentice-Hall International, Inc. NJ 07458,USA, 1997.

[10] M. Sugeno, and G.T. Kang, Structure Identification of Fuzzy Model, Fuzzy Sets and Systems, 28, 1988, pp. 15-33.

[11] Takagi, and M. Sugeno , "Fuzzy Identification of Systems and its Applications to Modeling and Control", IEEE Trans. On System, Man, and Cybernetics, Vol. 15, 1985 , pp. 116-132.

[12] Jeong-Woo Choi, Seung-Mok Ob, Hyun-Goo Choi, SangBaek Lee, Kwang-Soon Lee and Won-Hong Lee Fuzzy Control of Ethanol Concentration for Emulsan Production in a Fed-Batch Cultivation of Acinetobacter calcoaceticus RAG-I Korean 1. Chern. Eng., 15(3), 310(1998)

[13] Min Oh and II Moon, Framework of Dynamic Simulation for Complex Chemical Processes, Korean J. Chern. Eng., 15(3), 231(1998)

[14] M. S. Mahmoud, M. F. Hassan and M. G. Darwish, "Large Scale Control Systems: Theories and Techniques", Marcel Dekker Inc., New York, 1985.

[15] D. D. Siljak, "Decentralized Control of Complex Systems", Academic Press, Boston, 1991.

[16] M. S. Mahmoud, "Computer-Operated Systems Control", Marcel Dekker Inc., New York, 1991.

[17] M. Jamshidi, "Large Scale Systems: Modeling, Control and Fuzzy Logic", Prentice-Hall, New Jersey, 1997.

[18] M. S. Mahmoud, S.Z. Eid and A .A. Abou-Elseoud, "A Real Time Expert System for Dynamical Processes", IEEE Transactions Systems, Man and Cybernetics, Vol. SMC-19, No. 5, September/October 1989, pp. 1101-1105.

[19] M. S. Mahmoud, S. Kotob, and A. A. Abou-Elseoud, "A Learning Rule-Based Control System", Journal of Information and Decision Technologies, Vol. 18, No. 1, January 1992, pp. 55-66.

[20] M. S. Mahmoud, A. A. Abou-Elseoud and S. Kotob, "Development of Expert Control Systems: A Pattern Classification and Recognition Approach", Journal of Intelligent and Robotic Systems, Vol. 5, No. 2, April 1992, pp. 129-146.

[21] L. X. Wang, "A Supervisory Controller for Fuzzy Control Systems that Guarantees", IEEE Trans. Automatic Control, Vol. 39, No. 9, September 1994, pp. 1845-1847.

[22] K. M. Passino, and S. Yurkovick, "Fuzzy Control", Addison Wesley, California, 1998.

[23] C. C. Lee, "Fuzzy Logic in Control Systems: Fuzzy Logic Controller, Parts I & II", IEEE Systems, Man and Cybernetics, Vol. 20, No. 2, March/April 1990, pp. 404-435.

[24] M. Sugeno, and G.T. Kang, Structure Identification of Fuzzy Model, Fuzzy Sets and Systems, 28, 1988, pp. 15-33.

[25] Takagi, and M. Sugeno , Fuzzy Identification of Systems and its Applications to Modelling and Control, IEEE Trans. On System, Man,and Cybernetics, Vol. 15, 1985 , pp. 116-132

[26] Suciu, C; Liliana; Dafinca; Kansara M Margineanu ,"Switching capacitor fuzzy controller for power factor correction on inductive circuit", Power Electronics Specialist Conference, 31st Annual, 2000, IEEE, vol. 2.pp 773- 777.

[27] K.H . Abdul-Rahman, S.M. Shahidehpour, "Application of Fuzzy sets to optimal reactive power planning with security constrains", TransacTion of Power Systems, 1994, vol. 9, No.2, pp 589-597.

[28] Yu Qin; ShanShan Du, "To design fuzzy and digital controller for a single phase power factor pre-regulator-genetic algorithm approach", Industrial Application Conference, Annual, 1997 IEEE, vol, 2, pp 791-796.

[29] Yu Qin; ShanShan Du, "Control of single phase power factor pre-regulator for on-line uninterruptible power supply using fuzzy logic control inference", Applied power electronics Conference and Exposition, 11th Annual, 1996, vol.2, pp 699- 702.

[30] Borges da Silva, L.E; Ferreira da silva; M; Lambert-Torres, G, "An intelligent hybrid active power factor compensator for power systems,Systems", Man and Cybernetcs, 1995. Intelligentvsystem for the 21st Century, /EEE, vol.2 pp 1367- 1371.

[31] Turan Gonen, "Electrical power distribution system engineering", McGraw-Hill Inc. NY, 1986.

[32] Navd R. Zargari, Yuan Xiao; Bin Wu, "A multilevel thyristor rectifier with improved power faclor", Transaction on Industrial Applications, 1997,vol. 33 pp 1208- 1213.

[33] Chung,H.S .H; Tam,E.P.W; Huni,S.Y.R. "Development of a Fuzzy Logic Controller for boost rectifier with acti ve power factor correction", Power Electronics Specialist Conference, 30th Annual, 1999, IEEE, vol. 1, pp 149- 154.

[34] Mohammed T. Bi na; David C. Hamill , "The Bootstrap Varible Inductance: A New FACTS Control Elemenl".1999,l EEE.

[35] Muhammad Harunur Rash id, "Power Electronics Circuit, Devices and Applications", Prentice Hall Inc, NJ, 1988.

[36] Wang, Li-Xin, "A Course In Fuzzy Systems And Control", Prentice-Hall, N. J., 1997.

[37] ADAMS, J. A. Human factors in engineering. New York, Macmillan Publishing Company, 1989.

[38] Dominic Cooper C.Psychol "Human Factors in Accidents", Institute of Quarrying, North Of England- CoalPro Seminar, Ramside Hall, Durham, UK. 12 March 2002

[39] PEYTON, R. X.. Construction safety practices. New York, Van 1991.

[40] Sii, H.S., Wang, J., Ruxton, T., Yang, J.B., Liu, J. "Fuzzy logic approaches to safety assessment in maritime engineering applications", Journal of Marine Engineering & Technology (A5): 45-58. 2004.

[41] KARTAM, N. A. and BOUZ, R. G.. "Fatalities and Injuries in Kuwaiti Construction Industry", Accident Analysis and Prevention, (30), 805-814, 1991.

[42] A. Kamal, S. M. Eid and M. S. Mahmoud, "Multi-Stage Clustering: An Efficient Technique in Socioeconomic Field Experiments", IEEE Transactions Systems, Man and Cybernetics, Vol. SMC-11, No. 12, December 1981, pp. 779-785.

[43] BEZDEK, J. C. & PAL K. S. "Fuzzy models for pattern recognition". New York: IEEE Press, 1992.

[44] WINDHAM, M. P. "Cluster validity for the fuzzy c-mean clustering algorithm", IEEE Transactions Pattern Analysis and Machine Intelligence, (PAMI-4), 357-363, 1982 .

[45] XUANLI LIISA XIE & GERARDO BENI. "A Validity Measure for Fuzzy Clustering". IEEE Transactions Pattern Analysis and Machine Intelligence, vol. (PAMI-13), no. 8, pp. 841-847, August 1991.

[46] The Mathworks, "Fuzzy Logic Toolbox", Boston,1999.

[47] John J. Granger and William D. Stevenson, Jr., "Power System Analysis", McGraw-Hill, Inc., New York, USA, 1994.

Chapter 5

New Results on Robust $\mathcal{H}_\infty$ Filter for Uncertain Fuzzy Descriptor Systems

Wudhichai Assawinchaichote*

Additional information is available at the end of the chapter

1. Introduction

The problem of filter design for descriptor systems system has been intensively studied by a number of researchers for the past three decades; see Ref.[1]-[6]. This is due not only to theoretical interest but also to the relevance of this topic in control engineering applications. Descriptor systems or so called singularly perturbed systems are dynamical systems with multiple time-scales. Descriptor systems often occur naturally due to the presence of small "parasitic" parameter, typically small time constants, masses, etc.

The main purpose of the singular perturbation approach to analysis and design is the alleviation of high dimensionality and ill-conditioning resulting from the interaction of slow and fast dynamics modes. The separation of states into slow and fast ones is a nontrivial modelling task demanding insight and ingenuity on the part of the analyst. In state space, such systems are commonly modelled using the mathematical framework of singular perturbations, with a small parameter, say ε, determining the degree of separation between the "slow" and "fast" modes of the system.

In the last few years, many researchers have studied the $\mathcal{H}_\infty$ filter design for a general class of linear descriptor systems. In Ref.[3], the authors have investigated the decomposition solution of $\mathcal{H}_\infty$ filter gain for singularly perturbed systems. The reduced-order $\mathcal{H}_\infty$ optimal filtering for system with slow and fast modes has been considered in Ref.[4]. Although many researchers have studied linear descriptor systems for many years, the $\mathcal{H}_\infty$ filtering design for nonlinear descriptor systems remains as an open research area. This is because, in general, nonlinear singularly perturbed systems can not be easily separated into slow and fast subsystems.

Fuzzy system theory enables us to utilize qualitative, linguistic information about a highly complex nonlinear system to construct a mathematical model for it. Recent studies show

*W. Assawinchaichote is with the Department of Electronic and Telecommnunication Engineering, King Mongkut's University of Technology Thonburi, 126 Prachautits Rd., Bangkok 10140, Thailand.

that a fuzzy linear model can be used to approximate global behaviors of a highly complex nonlinear system; see for example, Ref.[7]-[19]. In this fuzzy linear model, local dynamics in different state space regions are represented by local linear systems. The overall model of the system is obtained by "blending" these linear models through nonlinear fuzzy membership functions. Unlike conventional modelling where a single model is used to describe the global behaviour of a system, the fuzzy modelling is essentially a multi-model approach in which simple sub-models (linear models) are combined to describe the global behaviour of the system.

What we intend to do in this paper is to design a robust $\mathcal{H}_\infty$ filter for a class of nonlinear descriptor systems with nonlinear on both fast and slow variables. First, we approximate this class of nonlinear descriptor systems by a Takagi-Sugeno fuzzy model. Then based on an LMI approach, we develop an $\mathcal{H}_\infty$ filter such that the $\mathcal{L}_2$-gain from an exogenous input to an estimate error is less or equal to a prescribed value. To alleviate the ill-conditioning resulting from the interaction of slow and fast dynamic modes, solutions to the problem are given in terms of linear matrix inequalities which are independent of the singular perturbation ε, when ε is sufficiently small. The proposed approach does not involve the separation of states into slow and fast ones and it can be applied not only to standard, but also to nonstandard nonlinear descriptor systems.

This paper is organized as follows. In Section 2, system descriptions and definitions are presented. In Section 3, based on an LMI approach, we develop a technique for designing a robust $\mathcal{H}_\infty$ filter for the system described in section 2. The validity of this approach is demonstrated by an example from a literature in Section 4. Finally in Section 5, conclusions are given.

2. System descriptions

In this section, we generalize the TS fuzzy system to represent a TS fuzzy descriptor system with parametric uncertainties. As in Ref.[19], we examine a TS fuzzy descriptor system with parametric uncertainties as follows:

$$
\begin{aligned}
E_\varepsilon \dot{x}(t) &= \sum_{i=1}^{r} \mu_i(v(t))\left[[A_i + \Delta A_i]x(t) + [B_{1_i} + \Delta B_{1_i}]w(t) + [B_{2_i} + \Delta B_{2_i}]u(t)\right] \\
z(t) &= \sum_{i=1}^{r} \mu_i(v(t))\left[[C_{1_i} + \Delta C_{1_i}]x(t) + [D_{12_i} + \Delta D_{12_i}]u(t)\right] \\
y(t) &= \sum_{i=1}^{r} \mu_i(v(t))\left[[C_{2_i} + \Delta C_{2_i}]x(t) + [D_{21_i} + \Delta D_{21_i}]w(t)\right]
\end{aligned}
\tag{1}
$$

where $E_\varepsilon = \begin{bmatrix} I & 0 \\ 0 & \varepsilon I \end{bmatrix}$, $\varepsilon > 0$ is the singular perturbation parameter, $v(t) = [v_1(t) \cdots v_\vartheta(t)]$ is the premise variable vector that may depend on states in many cases, $\mu_i(v(t))$ denotes the normalized time-varying fuzzy weighting functions for each rule (i.e., $\mu_i(v(t)) \geq 0$ and $\sum_{i=1}^{r} \mu_i(v(t)) = 1$), ϑ is the number of fuzzy sets, $x(t) \in \Re^n$ is the state vector, $u(t) \in \Re^m$ is the input, $w(t) \in \Re^p$ is the disturbance which belongs to $\mathcal{L}_2[0, \infty)$, $y(t) \in \Re^\ell$ is the measurement and $z(t) \in \Re^s$ is the controlled output, the matrices $A_i, B_{1_i}, B_{2_i}, C_{1_i}, C_{2_i}, D_{12_i}$ and D_{21_i} are of appropriate dimensions, and the matrices $\Delta A_i, \Delta B_{1_i}, \Delta B_{2_i}, \Delta C_{1_i}, \Delta C_{2_i}, \Delta D_{12_i}$ and ΔD_{21_i} represent the uncertainties in the system and satisfy the following assumption.

Assumption 1.

$$\Delta A_i = F(x(t),t)H_{1_i}, \quad \Delta B_{1_i} = F(x(t),t)H_{2_i}, \quad \Delta B_{2_i} = F(x(t),t)H_{3_i},$$

$$\Delta C_{1_i} = F(x(t),t)H_{4_i}, \quad \Delta C_{2_i} = F(x(t),t)H_{5_i}, \quad \Delta D_{12_i} = F(x(t),t)H_{6_i}$$

$$and \ \Delta D_{21_i} = F(x(t),t)H_{7_i}$$

where H_{j_i}, $j = 1,2,\cdots,7$ are known matrix functions which characterize the structure of the uncertainties. Furthermore, the following inequality holds:

$$\|F(x(t),t)\| \le \rho \tag{2}$$

for any known positive constant ρ.

Next, let us recall the following definition.

Definition 1. Suppose γ is a given positive number. A system (1) is said to have an $\mathcal{L}_2$-gain less than or equal to γ if

$$\int_0^{T_f} \left(z(t) - \hat{z}(t)\right)^T \left(z(t) - \hat{z}(t)\right) dt \le \gamma^2 \left[\int_0^{T_f} w^T(t)w(t)dt\right] \tag{3}$$

with $x(0) = 0$, where $(z(t) - \hat{z}(t))$ is the estimated error output, for all $T_f \ge 0$ and $w(t) \in \mathcal{L}_2[0,T_f]$.

3. Robust $\mathcal{H}_\infty$ fuzzy filter design

Without loss of generality, in this section, we assume that $u(t) = 0$. Let us recall the system (1) with $u(t) = 0$ as follows:

$$\begin{aligned}
E_\varepsilon \dot{x}(t) &= \sum_{i=1}^r \mu_i \left[[A_i + \Delta A_i]x(t) + [B_{1_i} + \Delta B_{1_i}]w(t)\right] \\
z(t) &= \sum_{i=1}^r \mu_i \left[[C_{1_i} + \Delta C_{1_i}]x(t)\right] \\
y(t) &= \sum_{i=1}^r \mu_i \left[[C_{2_i} + \Delta C_{2_i}]x(t) + [D_{21_i} + \Delta D_{21_i}]w(t)\right].
\end{aligned} \tag{4}$$

We are now aiming to design a full order dynamic $\mathcal{H}_\infty$ fuzzy filter of the form

$$\begin{aligned}
E_\varepsilon \dot{\hat{x}}(t) &= \sum_{i=1}^r \sum_{j=1}^r \hat{\mu}_i \hat{\mu}_j \left[\hat{A}_{ij}(\varepsilon)\hat{x}(t) + \hat{B}_i y(t)\right] \\
\hat{z}(t) &= \sum_{i=1}^r \hat{\mu}_i \hat{C}_i \hat{x}(t)
\end{aligned} \tag{5}$$

where $\hat{x}(t) \in \mathfrak{R}^n$ is the filter's state vector, $\hat{z} \in \mathfrak{R}^s$ is the estimate of $z(t)$, $\hat{A}_{ij}(\varepsilon)$, $\hat{B}_i$ and $\hat{C}_i$ are parameters of the filter which are to be determined, and $\hat{\mu}_i$ denotes the normalized time-varying fuzzy weighting functions for each rule (i.e., $\hat{\mu}_i \ge 0$ and $\sum_{i=1}^r \hat{\mu}_i = 1$), such that the inequality (3) holds. Clearly, in real control problems, all of the premise variables are not necessarily measurable. In this section, we then consider the designing of the robust $\mathcal{H}_\infty$ fuzzy filter into two cases as follows.

3.1. Case I–$v(t)$ is available for feedback

The premise variable of the fuzzy model $v(t)$ is available for feedback which implies that μ_i is available for feedback. Thus, we can select our filter that depends on μ_i as follows:

$$\begin{aligned}
E_\varepsilon \dot{\hat{x}}(t) &= \sum_{i=1}^{r}\sum_{j=1}^{r}\mu_i\mu_j\left[\hat{A}_{ij}(\varepsilon)\hat{x}(t)+\hat{B}_iy(t)\right]\\
\hat{z}(t) &= \sum_{i=1}^{r}\mu_i\hat{C}_i\hat{x}(t).
\end{aligned}\tag{6}$$

Before presenting our next results, the following lemma is recalled.

Lemma 1. *Consider the system (4). Given a prescribed $\mathcal{H}_\infty$ performance $\gamma > 0$ and a positive constant δ, if there exist matrices $X_\varepsilon = X_\varepsilon^T$, $Y_\varepsilon = Y_\varepsilon^T$, $\mathcal{B}_i(\varepsilon)$ and $C_i(\varepsilon)$, $i = 1,2,\cdots,r$, satisfying the following ε-dependent linear matrix inequalities:*

$$\begin{bmatrix} X_\varepsilon & I \\ I & Y_\varepsilon \end{bmatrix} > 0 \tag{7}$$

$$X_\varepsilon > 0 \tag{8}$$

$$Y_\varepsilon > 0 \tag{9}$$

$$\Psi_{11_{ii}}(\varepsilon) < 0, \quad i = 1,2,\cdots,r \tag{10}$$

$$\Psi_{22_{ii}}(\varepsilon) < 0, \quad i = 1,2,\cdots,r \tag{11}$$

$$\Psi_{11_{ij}}(\varepsilon)+\Psi_{11_{ji}}(\varepsilon) < 0, \quad i < j \leq r \tag{12}$$

$$\Psi_{22_{ij}}(\varepsilon)+\Psi_{22_{ji}}(\varepsilon) < 0, \quad i < j \leq r \tag{13}$$

where

$$\Psi_{11_{ij}}(\varepsilon) = \begin{pmatrix} E_\varepsilon^{-1}A_iY_\varepsilon + Y_\varepsilon A_i^T E_\varepsilon^{-1} + \gamma^{-2}E_\varepsilon^{-1}\breve{B}_{1_j}\breve{B}_{1_j}^T E_\varepsilon^{-1} & (*)^T \\ \left[Y_\varepsilon\breve{C}_{1_i}^T + E_\varepsilon^{-1}C_i^T(\varepsilon)\breve{D}_{12}^T\right]^T & -I \end{pmatrix} \tag{14}$$

$$\Psi_{22_{ij}}(\varepsilon) = \begin{pmatrix} A_i^T E_\varepsilon^{-1}X_\varepsilon + X_\varepsilon E_\varepsilon^{-1}A_i + \mathcal{B}_i(\varepsilon)C_{2_j} + C_{2_j}^T\mathcal{B}_j^T(\varepsilon) + \breve{C}_{1_i}^T\breve{C}_{1_i} & (*)^T \\ \left[X_\varepsilon E_\varepsilon^{-1}\breve{B}_{1_i} + \mathcal{B}_i(\varepsilon)\breve{D}_{21_j}\right]^T & -\gamma^2 I \end{pmatrix} \tag{15}$$

with

$$\breve{B}_{1_i} = \begin{bmatrix} \delta I & I & 0 & B_{1_i} & 0 \end{bmatrix},$$

$$\breve{C}_{1_i} = \begin{bmatrix} \frac{\gamma\rho}{\delta}H_{1_i}^T & \frac{\gamma\rho}{\delta}H_{5_i}^T & \sqrt{2}\lambda\rho H_{4_i}^T & \sqrt{2}\lambda C_{1_i}^T \end{bmatrix}^T,$$

$$\breve{D}_{12} = \begin{bmatrix} 0 & 0 & 0 & -\sqrt{2}\lambda I \end{bmatrix}^T,$$

$$\breve{D}_{21_i} = \begin{bmatrix} 0 & 0 & \delta I & D_{21_i} & I \end{bmatrix}$$

and $\lambda = \left(1+\rho^2\sum_{i=1}^{r}\sum_{j=1}^{r}\left[\|H_{2_i}^T H_{2_j}\| + \|H_{7_i}^T H_{7_j}\|\right]\right)^{\frac{1}{2}},$

then the prescribed $\mathcal{H}_\infty$ performance $\gamma > 0$ is guaranteed. Furthermore, a suitable filter is of the form (6) with

$$\begin{aligned}
\hat{A}_{ij}(\varepsilon) &= E_\varepsilon[Y_\varepsilon^{-1}-X_\varepsilon]^{-1}M_{ij}(\varepsilon)Y_\varepsilon^{-1}\\
\hat{B}_i &= E_\varepsilon[Y_\varepsilon^{-1}-X_\varepsilon]^{-1}\mathcal{B}_i(\varepsilon)\\
\hat{C}_i &= C_i(\varepsilon)E_\varepsilon^{-1}Y_\varepsilon^{-1}
\end{aligned}\tag{16}$$

where

$$\mathcal{M}_{ij}(\varepsilon) = -A_i^T E_\varepsilon^{-1} - X_\varepsilon E_\varepsilon^{-1} A_i Y_\varepsilon - \left[Y_\varepsilon^{-1} - X_\varepsilon\right] E_\varepsilon^{-1} \hat{B}_i C_{2_j} Y_\varepsilon - \tilde{C}_{1_i}^T \left[\tilde{C}_{1_j} Y_\varepsilon + \tilde{D}_{12} \hat{C}_j Y_\varepsilon\right]$$
$$-\gamma^{-2} \left\{ X_\varepsilon E_\varepsilon^{-1} \tilde{B}_{1_i} + \left[Y_\varepsilon^{-1} - X_\varepsilon\right] E_\varepsilon^{-1} \hat{B}_i \tilde{D}_{21_i} \right\} \tilde{B}_{1_j}^T E_\varepsilon^{-1}.$$

Proof: It can be shown by employing the same technique used in Ref.[18]-[19]. ∎

Remark 1. *The LMIs given in Lemma 1 may become ill-conditioned when ε is sufficiently small, which is always the case for the descriptor systems. In general, these ill-conditioned LMIs are very difficult to solve. Thus, to alleviate these ill-conditioned LMIs, we have the following ε-independent well-posed LMI-based sufficient conditions for the uncertain fuzzy descriptor systems to obtain the prescribed $\mathcal{H}_\infty$ performance.* □

Theorem 1. *Consider the system (4). Given a prescribed $\mathcal{H}_\infty$ performance $\gamma > 0$ and a positive constant δ, if there exist matrices X_0, Y_0, $\mathcal{B}_{0_i}$ and $\mathcal{C}_{0_i}$, $i = 1, 2, \cdots, r$, satisfying the following ε-independent linear matrix inequalities:*

$$\begin{bmatrix} X_0 E + DX_0 & I \\ I & Y_0 E + DY_0 \end{bmatrix} > 0 \tag{17}$$

$$EX_0^T = X_0 E, \quad X_0^T D = DX_0, \quad X_0 E + DX_0 > 0 \tag{18}$$

$$EY_0^T = Y_0 E, \quad Y_0^T D = DY_0, \quad Y_0 E + DY_0 > 0 \tag{19}$$

$$\Psi_{11_{ii}} < 0, \quad i = 1, 2, \cdots, r \tag{20}$$

$$\Psi_{22_{ii}} < 0, \quad i = 1, 2, \cdots, r \tag{21}$$

$$\Psi_{11_{ij}} + \Psi_{11_{ji}} < 0, \quad i < j \le r \tag{22}$$

$$\Psi_{22_{ij}} + \Psi_{22_{ji}} < 0, \quad i < j \le r \tag{23}$$

where $E = \begin{pmatrix} I & 0 \\ 0 & 0 \end{pmatrix}$, $D = \begin{pmatrix} 0 & 0 \\ 0 & I \end{pmatrix}$,

$$\Psi_{11_{ij}} = \begin{pmatrix} A_i Y_0^T + Y_0 A_i^T + \gamma^{-2} \tilde{B}_{1_i} \tilde{B}_{1_j}^T & (*)^T \\ \left[Y_0 \tilde{C}_{1_i}^T + C_{0_i}^T \tilde{D}_{12}^T\right]^T & -I \end{pmatrix} \tag{24}$$

$$\Psi_{22_{ij}} = \begin{pmatrix} A_i^T X_0^T + X_0 A_i + \mathcal{B}_{0_i} C_{2_j} + C_{2_i}^T \mathcal{B}_{0_j}^T + \tilde{C}_{1_i}^T \tilde{C}_{1_j} & (*)^T \\ \left[X_0 \tilde{B}_{1_i} + \mathcal{B}_{0_i} \tilde{D}_{21_j}\right]^T & -\gamma^2 I \end{pmatrix} \tag{25}$$

with

$$\tilde{B}_{1_i} = \begin{bmatrix} \delta I & I & 0 & B_{1_i} & 0 \end{bmatrix},$$

$$\tilde{C}_{1_i} = \begin{bmatrix} \frac{\gamma \rho}{\delta} H_{1_i}^T & \frac{\gamma \rho}{\delta} H_{5_i}^T & \sqrt{2}\lambda \rho H_{4_i}^T & \sqrt{2}\lambda C_{1_i}^T \end{bmatrix}^T,$$

$$\tilde{D}_{12} = \begin{bmatrix} 0 & 0 & 0 & -\sqrt{2}\lambda I \end{bmatrix}^T,$$

$$\tilde{D}_{21_i} = \begin{bmatrix} 0 & 0 & \delta I & D_{21_i} & I \end{bmatrix}$$

and $\lambda = \left(1 + \rho^2 \sum_{i=1}^r \sum_{j=1}^r \left[\|H_{2_i}^T H_{2_j}\| + \|H_{7_i}^T H_{7_j}\|\right]\right)^{\frac{1}{2}}$, *then there exists a sufficiently small $\hat{\varepsilon} > 0$ such that for $\varepsilon \in (0, \hat{\varepsilon}]$, the prescribed $\mathcal{H}_\infty$ performance $\gamma > 0$ is guaranteed. Furthermore, a suitable filter is of the form (6) with*

$$\hat{A}_{ij}(\varepsilon) = \left[Y_\varepsilon^{-1} - X_\varepsilon\right]^{-1} \mathcal{M}_{0_{ij}}(\varepsilon) Y_\varepsilon^{-1}$$
$$\hat{B}_i \quad = \left[Y_0^{-1} - X_0\right]^{-1} B_{0_i} \tag{26}$$
$$\hat{C}_i \quad = C_{0_i} Y_0^{-1}$$

where

$$\mathcal{M}_{0_{ij}}(\varepsilon) = -A_i^T - X_\varepsilon A_i Y_\varepsilon - \left[Y_\varepsilon^{-1} - X_\varepsilon\right]\hat{B}_i C_{2_j} Y_\varepsilon - \tilde{C}_{1_i}^T \left[\tilde{C}_{1_j} Y_\varepsilon + \tilde{D}_{12}\hat{C}_j Y_\varepsilon\right]$$

$$-\gamma^{-2}\left\{X_\varepsilon \tilde{B}_{1_i} + \left[Y_\varepsilon^{-1} - X_\varepsilon\right]\hat{B}_i \tilde{D}_{21_i}\right\}\tilde{B}_{1_j}^T$$

$$X_\varepsilon = \left\{X_0 + \varepsilon\tilde{X}\right\}E_\varepsilon \text{ and } Y_\varepsilon^{-1} = \left\{Y_0^{-1} + \varepsilon N_\varepsilon\right\}E_\varepsilon \tag{27}$$

with $\tilde{X} = D\left(X_0^T - X_0\right)$ and $N_\varepsilon = D\left((Y_0^{-1})^T - Y_0^{-1}\right)$.

Proof: Suppose the inequalities (17)-(19) hold, then the matrices X_0 and Y_0 are of the following forms:

$$X_0 = \begin{pmatrix} X_1 & X_2 \\ 0 & X_3 \end{pmatrix} \text{ and } Y_0 = \begin{pmatrix} Y_1 & Y_2 \\ 0 & Y_3 \end{pmatrix}$$

with $X_1 = X_1^T > 0$, $X_3 = X_3^T > 0$, $Y_1 = Y_1^T > 0$ and $Y_3 = Y_3^T > 0$. Substituting X_0 and Y_0 into (27), respectively, we have

$$X_\varepsilon = \left\{X_0 + \varepsilon\tilde{X}\right\}E_\varepsilon = \begin{pmatrix} X_1 & \varepsilon X_2 \\ \varepsilon X_2^T & \varepsilon X_3 \end{pmatrix} \tag{28}$$

$$Y_\varepsilon^{-1} = \left\{Y_0^{-1} + \varepsilon N_\varepsilon\right\}E_\varepsilon = \begin{pmatrix} Y_1^{-1} & -\varepsilon Y^{-1}Y_2 Y_3^{-1} \\ -\varepsilon(Y^{-1}Y_2 Y_3^{-1})^T & \varepsilon Y_3^{-1} \end{pmatrix}. \tag{29}$$

Clearly, $X_\varepsilon = X_\varepsilon^T$, and $Y_\varepsilon^{-1} = (Y_\varepsilon^{-1})^T$. Knowing the fact that the inverse of a symmetric matrix is a symmetric matrix, we learn that Y_ε is a symmetric matrix. Using the matrix inversion lemma, we can see that

$$Y_\varepsilon = E_\varepsilon^{-1}\left\{Y_0 + \varepsilon\tilde{Y}\right\} \tag{30}$$

where $\tilde{Y} = Y_0 N_\varepsilon (I + \varepsilon Y_0 N_\varepsilon)^{-1} Y_0$. Employing the Schur complement, one can show that there exists a sufficiently small $\hat{\varepsilon}$ such that for $\varepsilon \in (0, \hat{\varepsilon}]$, (8)-(9) holds.

Now, we need to show that

$$\begin{pmatrix} X_\varepsilon & I \\ I & Y_\varepsilon \end{pmatrix} > 0. \tag{31}$$

By the Schur complement, it is equivalent to showing that

$$X_\varepsilon - Y_\varepsilon^{-1} > 0. \tag{32}$$

Substituting (28) and (29) into the left hand side of (32), we get

$$\begin{bmatrix} X_1 - Y_1^{-1} & \varepsilon(X_2 + Y_1^{-1}Y_2Y_3^{-1}) \\ \varepsilon(X_2 + Y_1^{-1}Y_2Y_3^{-1})^T & \varepsilon(X_3 - Y_3^{-1}) \end{bmatrix}. \tag{33}$$

The Schur complement of (17) is

$$\begin{bmatrix} X_1 - Y_1^{-1} & 0 \\ 0 & X_3 - Y_3^{-1} \end{bmatrix} > 0. \tag{34}$$

According to (34), we learn that

$$X_1 - Y_1^{-1} > 0 \quad \text{and} \quad X_3 - Y_3^{-1} > 0. \tag{35}$$

Using (35) and the Schur complement, it can be shown that there exists a sufficiently small $\hat{\varepsilon} > 0$ such that for $\varepsilon \in (0, \hat{\varepsilon}]$, (7) holds.

Next, employing (28), (29) and (30), the controller's matrices given in (16) can be re-expressed as follows:

$$\mathcal{B}_i(\varepsilon) = [Y_0^{-1} - X_0]\hat{B}_i + \varepsilon[N_\varepsilon - \tilde{X}]\hat{B}_i \triangleq \mathcal{B}_{0_i} + \varepsilon\mathcal{B}_{\varepsilon_i}$$

$$\mathcal{C}_i(\varepsilon) = \hat{C}_i Y_0^T + \varepsilon\hat{C}_i \tilde{Y}^T \triangleq \mathcal{C}_{0_i} + \varepsilon\mathcal{C}_{\varepsilon_i}. \tag{36}$$

Substituting (28), (29), (30) and (36) into (14) and (15), and pre-post multiplying by $\begin{pmatrix} E_\varepsilon & 0 \\ 0 & I \end{pmatrix}$, we, respectively, obtain

$$\Psi_{11_{ij}} + \psi_{11_{ij}} \quad \text{and} \quad \Psi_{22_{ij}} + \psi_{22_{ij}} \tag{37}$$

where the ε-independent linear matrices $\Psi_{11_{ij}}$ and $\Psi_{22_{ij}}$ are defined in (24) and (25), respectively and the ε-dependent linear matrices are

$$\psi_{11_{ij}} = \varepsilon \begin{pmatrix} A_i\tilde{Y}^T + \tilde{Y}A_i^T & (*)^T \\ \left[\tilde{Y}\tilde{C}_{1_i}^T + \mathcal{C}_{\varepsilon_i}^T\tilde{D}_{12_j}^T\right]^T & 0 \end{pmatrix} \tag{38}$$

$$\psi_{22_{ij}} = \varepsilon \begin{pmatrix} A_i^T\tilde{X} + \tilde{X}^T A_i + \mathcal{B}_{\varepsilon_i}C_{2_j} + C_{2_j}^T\mathcal{B}_{\varepsilon_i}^T & (*)^T \\ \left[\tilde{X}\tilde{B}_{1_i} + \mathcal{B}_{\varepsilon_i}\tilde{D}_{21_j}\right]^T & 0 \end{pmatrix}. \tag{39}$$

Note that the ε-dependent linear matrices tend to zero when ε approaches zero.

Employing (20)-(22) and knowing the fact that for any given negative definite matrix $\mathcal{W}$, there exists an $\varepsilon > 0$ such that $\mathcal{W} + \varepsilon I < 0$, one can show that there exists a sufficiently small $\hat{\varepsilon} > 0$ such that for $\varepsilon \in (0, \hat{\varepsilon}]$, (10)-(13) hold. Since (7)-(13) hold, using Lemma 1, the inequality (3) holds. ∎

3.2. Case II–$v(t)$ is unavailable for feedback

The fuzzy filter is assumed to be the same as the premise variables of the fuzzy system model. This actually means that the premise variables of fuzzy system model are assumed to be measurable. However, in general, it is extremely difficult to derive an accurate fuzzy system model by imposing that all premise variables are measurable. In this subsection, we do not impose that condition, we choose the premise variables of the filter to be different from the premise variables of fuzzy system model of the plant. In here, the premise variables of the filter are selected to be the estimated premise variables of the plant. In the other words, the premise variable of the fuzzy model $v(t)$ is unavailable for feedback which implies μ_i is unavailable for feedback. Hence, we cannot select our filter which depends on μ_i. Thus, we select our filter as (5) where $\hat{\mu}_i$ depends on the premise variable of the filter which is different from μ_i. Let us re-express the system (1) in terms of $\hat{\mu}_i$, thus the plant's premise variable becomes the same as the filter's premise variable. By doing so, the result given in the previous case can then be applied here. Note that it can be done by using the same technique as in subsection. After some manipulation, we get

$$\begin{aligned}
E_\varepsilon \dot{x}(t) &= \sum_{i=1}^r \hat{\mu}_i \Big[[A_i + \Delta\bar{A}_i]x(t) + [B_{1_i} + \Delta\bar{B}_{1_i}]w(t) \\
z(t) &= \sum_{i=1}^r \hat{\mu}_i \Big[[C_{1_i} + \Delta\bar{C}_{1_i}]x(t)\Big] \\
y(t) &= \sum_{i=1}^r \hat{\mu}_i \Big[[C_{2_i} + \Delta\bar{C}_{2_i}]x(t) + [D_{21_i} + \Delta\bar{D}_{21_i}]w(t)\Big]
\end{aligned} \tag{40}$$

where

$$\Delta\bar{A}_i = \bar{F}(x(t),\hat{x}(t),t)\bar{H}_{1_i}, \quad \Delta\bar{B}_{1_i} = \bar{F}(x(t),\hat{x}(t),t)\bar{H}_{2_i}, \quad \Delta\bar{B}_{2_i} = \bar{F}(x(t),\hat{x}(t),t)\bar{H}_{3_i},$$

$$\Delta\bar{C}_{1_i} = \bar{F}(x(t),\hat{x}(t),t)\bar{H}_{4_i}, \quad \Delta\bar{C}_{2_i} = \bar{F}(x(t),\hat{x}(t),t)\bar{H}_{5_i}, \quad \Delta\bar{D}_{12_i} = \bar{F}(x(t),\hat{x}(t),t)\bar{H}_{6_i}$$

$$\text{and} \quad \Delta\bar{D}_{21_i} = \bar{F}(x(t),\hat{x}(t),t)\bar{H}_{7_i}$$

with

$$\bar{H}_{1_i} = \Big[H_{1_i}^T\ A_1^T\ \cdots\ A_r^T\ H_{1_1}^T\ \cdots\ H_{1_r}^T\Big]^T, \quad \bar{H}_{2_i} = \Big[H_{2_i}^T\ B_{1_1}^T\ \cdots\ B_{1_r}^T\ H_{2_1}^T\ \cdots\ H_{2_r}^T\Big]^T,$$

$$\bar{H}_{3_i} = \Big[H_{3_i}^T\ B_{2_1}^T\ \cdots\ B_{2_r}^T\ H_{3_1}^T\ \cdots\ H_{3_r}^T\Big]^T, \quad \bar{H}_{4_i} = \Big[H_{4_i}^T\ C_{1_1}^T\ \cdots\ C_{1_r}^T\ H_{4_1}^T\ \cdots\ H_{4_r}^T\Big]^T,$$

$$\bar{H}_{5_i} = \Big[H_{5_i}^T\ C_{2_1}^T\ \cdots\ C_{2_r}^T\ H_{5_1}^T\ \cdots\ H_{5_r}^T\Big]^T, \quad \bar{H}_{6_i} = \Big[H_{6_i}^T\ D_{12_1}^T\ \cdots\ D_{12_r}^T\ H_{6_1}^T\ \cdots\ H_{6_r}^T\Big]^T$$

$$\bar{H}_{7_i} = \Big[H_{7_i}^T\ D_{21_1}^T\ \cdots\ D_{21_r}^T\ H_{7_1}^T\ \cdots\ H_{7_r}^T\Big]^T \quad \text{and}$$

$\bar{F}(x(t),\hat{x}(t),t) = \Big[F(x(t),t)\ (\mu_1 - \hat{\mu}_1)\ \cdots\ (\mu_r - \hat{\mu}_r)\ F(x(t),t)(\mu_1 - \hat{\mu}_1)\ \cdots\ F(x(t),t)(\mu_r - \hat{\mu}_r)\Big]$. Note that $\|\bar{F}(x(t),\hat{x}(t),t)\| \leq \bar{\rho}$ where $\bar{\rho} = \{3\rho^2 + 2\}^{\frac{1}{2}}$. $\bar{\rho}$ is derived by utilizing the concept of vector norm in the basic system control theory and the fact that $\mu_i \geq 0$, $\hat{\mu}_i \geq 0$, $\sum_{i=1}^r \mu_i = 1$ and $\sum_{i=1}^r \hat{\mu}_i = 1$.

Note that the above technique is basically employed in order to obtain the plant's premise variable to be the same as the filter's premise variable; e.g. [17]. Now, the premise variable of the system is the same as the premise variable of the filter, thus we can apply the result given in Case I. By applying the same technique used in Case I, we have the following theorem.

Theorem 2. *Consider the system* (4). *Given a prescribed* $\mathcal{H}_\infty$ *performance* $\gamma > 0$ *and a positive constant* δ, *if there exist matrices* X_0, Y_0, $\mathcal{B}_{0_i}$ *and* $\mathcal{C}_{0_i}$, $i = 1, 2, \cdots, r$, *satisfying the following* ε-*independent linear matrix inequalities:*

$$\begin{bmatrix} X_0 E + D X_0 & I \\ I & Y_0 E + D Y_0 \end{bmatrix} \not> 0 \tag{41}$$

$$E X_0^T = X_0 E, \quad X_0^T D = D X_0, \quad X_0 E + D X_0 \not> 0 \tag{42}$$

$$E Y_0^T = Y_0 E, \quad Y_0^T D = D Y_0, \quad Y_0 E + D Y_0 \not> 0 \tag{43}$$

$$\Psi_{11_{ii}} < 0, \quad i = 1, 2, \cdots, r \tag{44}$$

$$\Psi_{22_{ii}} < 0, \quad i = 1, 2, \cdots, r \tag{45}$$

$$\Psi_{11_{ij}} + \Psi_{11_{ji}} < 0, \quad i < j \leq r \tag{46}$$

$$\Psi_{22_{ij}} + \Psi_{22_{ji}} < 0, \quad i < j \leq r \tag{47}$$

where $E = \begin{pmatrix} I & 0 \\ 0 & 0 \end{pmatrix}, D = \begin{pmatrix} 0 & 0 \\ 0 & I \end{pmatrix},$

$$\Psi_{11_{ij}} = \begin{pmatrix} A_i Y_0^T + Y_0 A_i^T + \gamma^{-2} \breve{\bar{B}}_{1_i} \breve{\bar{B}}_{1_i}^T & (*)^T \\ [Y_0 \breve{\bar{C}}_{1_i}^T + C_{0_i}^T \breve{\bar{D}}_{12}^T]^T & -I \end{pmatrix}$$

$$\Psi_{22_{ij}} = \begin{pmatrix} A_i^T X_0^T + X_0 A_i + \mathcal{B}_{0_i} C_{2_j} + C_{2_j}^T \mathcal{B}_{0_j}^T + \breve{\bar{C}}_{1_i}^T \breve{\bar{C}}_{1_i} & (*)^T \\ [X_0 \breve{\bar{B}}_{1_i} + \mathcal{B}_{0_i} \breve{\bar{D}}_{21_j}]^T & -\gamma^2 I \end{pmatrix}$$

with

$$\breve{\bar{B}}_{1_i} = \begin{bmatrix} \delta I & I & 0 & B_{1_i} & 0 \end{bmatrix},$$

$$\breve{\bar{C}}_{1_i} = \begin{bmatrix} \frac{\gamma \bar{\rho}}{\delta} \bar{H}_{1_i}^T & \frac{\gamma \bar{\rho}}{\delta} \bar{H}_{5_i}^T & \sqrt{2} \bar{\lambda} \bar{\rho} \bar{H}_{4_i}^T & \sqrt{2} \bar{\lambda} C_{1_i}^T \end{bmatrix}^T,$$

$$\breve{\bar{D}}_{12} = \begin{bmatrix} 0 & 0 & 0 & -\sqrt{2} \bar{\lambda} I \end{bmatrix}^T,$$

$$\breve{\bar{D}}_{21_i} = \begin{bmatrix} 0 & 0 & \delta I & D_{21_i} & I \end{bmatrix}$$

and $\bar{\lambda} = \left(1 + \bar{\rho}^2 \sum_{i=1}^{r} \sum_{j=1}^{r} \left[\| \bar{H}_{2_i}^T \bar{H}_{2_j} \| + \| \bar{H}_{7_i}^T \bar{H}_{7_j} \| \right] \right)^{\frac{1}{2}},$

then there exists a sufficiently small $\hat{\varepsilon} > 0$ *such that for* $\varepsilon \in (0, \hat{\varepsilon}]$, *the prescribed* $\mathcal{H}_\infty$ *performance* $\gamma > 0$ *is guaranteed. Furthermore, a suitable filter is of the form* (**??**) *with*

$$\begin{aligned} \hat{A}_{ij}(\varepsilon) &= \left[Y_\varepsilon^{-1} - X_\varepsilon \right]^{-1} \mathcal{M}_{0_{ij}}(\varepsilon) Y_\varepsilon^{-1} \\ \hat{B}_i &= \left[Y_0^{-1} - X_0 \right]^{-1} \mathcal{B}_{0_i} \\ \hat{C}_i &= \mathcal{C}_{0_i} Y_0^{-1} \end{aligned} \tag{48}$$

where

$$\mathcal{M}_{0_{ij}}(\varepsilon) = -A_i^T - X_\varepsilon A_i Y_\varepsilon - \left[Y_\varepsilon^{-1} - X_\varepsilon \right] \hat{B}_i C_{2_j} Y_\varepsilon - \breve{\bar{C}}_{1_i}^T \left[\breve{\bar{C}}_{1_j} Y_\varepsilon + \breve{\bar{D}}_{12} \hat{C}_j Y_\varepsilon \right]$$

$$-\gamma^{-2} \left\{ X_\varepsilon \breve{\bar{B}}_{1_i} + \left[Y_\varepsilon^{-1} - X_\varepsilon \right] \hat{B}_i \breve{\bar{D}}_{21_j} \right\} \breve{\bar{B}}_{1_i}^T$$

$$X_\varepsilon = \left\{ X_0 + \varepsilon \tilde{X} \right\} E_\varepsilon \text{ and } Y_\varepsilon^{-1} = \left\{ Y_0^{-1} + \varepsilon N_\varepsilon \right\} E_\varepsilon$$

with $\tilde{X} = D \left(X_0^T - X_0 \right)$ *and* $N_\varepsilon = D \left((Y_0^{-1})^T - Y_0^{-1} \right).$

Proof: It can be shown by employing the same technique used in the proof for Theorem 1. ∎

4. Example

Consider the tunnel diode circuit shown in Figure 1 where the tunnel diode is characterized by

$$i_D(t) = 0.01v_D(t) + 0.05v_D^3(t).$$

Assuming that the inductance, L, is the parasitic parameter and letting $x_1(t) = v_C(t)$ and

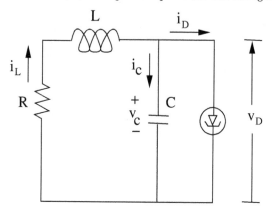

Figure 1. Tunnel diode circuit.

$x_2(t) = i_L(t)$ as the state variables, we have

$$\begin{aligned}
C\dot{x}_1(t) &= -0.01x_1(t) - 0.05x_1^3(t) + x_2(t) \\
L\dot{x}_2(t) &= -x_1(t) - Rx_2(t) + 0.1w_2(t) \\
y(t) &= Jx(t) + 0.1w_1(t) \\
z(t) &= \begin{bmatrix} x_1(t) \\ x_2(t) \end{bmatrix}
\end{aligned} \tag{49}$$

where $w(t)$ is the disturbance noise input, $y(t)$ is the measurement output, $z(t)$ is the state to be estimated and J is the sensor matrix. Note that the variables $x_1(t)$ and $x_2(t)$ are treated as the deviation variables (variables deviate from the desired trajectories). The parameters of the circuit are $C = 100 \ mF$, $R = 10 \pm 10\% \ \Omega$ and $L = \varepsilon \ H$. With these parameters (49) can be rewritten as

$$\begin{aligned}
\dot{x}_1(t) &= -0.1x_1(t) + 0.5x_1^3(t) + 10x_2(t) \\
\varepsilon\dot{x}_2(t) &= -x_1(t) - (10 + \Delta R)x_2(t) + 0.1w_2(t) \\
y(t) &= Jx(t) + 0.1w_1(t) \\
z(t) &= \begin{bmatrix} x_1(t) \\ x_2(t) \end{bmatrix}.
\end{aligned} \tag{50}$$

For the sake of simplicity, we will use as few rules as possible. Assuming that $|x_1(t)| \leq 3$, the nonlinear network system (50) can be approximated by the following TS fuzzy model:

Plant Rule 1: IF $x_1(t)$ is $M_1(x_1(t))$ THEN

$$E_\varepsilon \dot{x}(t) = [A_1 + \Delta A_1]x(t) + B_{1_1} w(t), \quad x(0) = 0,$$
$$z(t) = C_{1_1} x(t),$$
$$y(t) = C_{2_1} x(t) + D_{21_1} w(t).$$

Plant Rule 2: IF $x_1(t)$ is $M_2(x_1(t))$ THEN

$$E_\varepsilon \dot{x}(t) = [A_2 + \Delta A_2]x(t) + B_{1_2} w(t), \quad x(0) = 0,$$
$$z(t) = C_{1_2} x(t),$$
$$y(t) = C_{2_2} x(t) + D_{21_2} w(t)$$

where $x(t) = [x_1^T(t) \ x_2^T(t)]^T$, $w(t) = [w_1^T(t) \ w_2^T(t)]^T$,

$$A_1 = \begin{bmatrix} -0.1 & 10 \\ -1 & -1 \end{bmatrix}, \quad A_2 = \begin{bmatrix} -4.6 & 10 \\ -1 & -1 \end{bmatrix}, \quad B_{1_1} = B_{1_2} = \begin{bmatrix} 0 & 0 \\ 0 & 0.1 \end{bmatrix},$$

$$C_1 = \begin{bmatrix} 1 & 0 \\ 0 & 1 \end{bmatrix}, \quad C_{2_1} = C_{2_2} = J, \quad D_{21} = \begin{bmatrix} 0.1 & 0 \end{bmatrix},$$

$$\Delta A_1 = F(x(t), t) H_{1_1}, \quad \Delta A_2 = F(x(t), t) H_{1_2} \quad \text{and} \quad E_\varepsilon = \begin{bmatrix} 1 & 0 \\ 0 & \varepsilon \end{bmatrix}.$$

Now, by assuming that $\|F(x(t), t)\| \le \rho = 1$ and since the values of R are uncertain but bounded within 10% of their nominal values given in (49), we have

$$H_{1_1} = H_{1_2} = \begin{bmatrix} 0 & 0 \\ 0 & 1 \end{bmatrix}.$$

Note that the plot of the membership function Rules 1 and 2 is the same as in Figure 2. By employing the results given in Lemma 1 and the Matlab LMI solver, it is easy to realize that $\varepsilon < 0.006$ for the fuzzy filter design in Case I and $\varepsilon < 0.008$ for the fuzzy filter design in Case II, the LMIs become ill-conditioned and the Matlab LMI solver yields the error message, "Rank Deficient". *Case I-$v(t)$ are available for feedback*

In this case, $x_1(t) = v(t)$ is assumed to be available for feedback; for instance, $J = [1 \ 0]$. This implies that μ_i is available for feedback. Using the LMI optimization algorithm and Theorem 1 with $\varepsilon = 100 \ \mu H$, $\gamma = 0.6$ and $\delta = 1$, we obtain the following results:

$$\hat{A}_{11}(\varepsilon) = \begin{bmatrix} -0.0674 & -0.3532 \\ -30.7181 & -4.3834 \end{bmatrix}, \qquad \hat{A}_{12}(\varepsilon) = \begin{bmatrix} -0.0674 & -0.3532 \\ -30.7181 & -4.3834 \end{bmatrix},$$

$$\hat{A}_{21}(\varepsilon) = \begin{bmatrix} -0.0928 & -0.3138 \\ -34.7355 & -3.8964 \end{bmatrix}, \qquad \hat{A}_{22}(\varepsilon) = \begin{bmatrix} -0.0928 & -0.3138 \\ -34.7355 & -3.8964 \end{bmatrix},$$

$$\hat{B}_1 = \begin{bmatrix} 1.5835 \\ 3.2008 \end{bmatrix}, \qquad\qquad \hat{B}_2 = \begin{bmatrix} 1.2567 \\ 3.8766 \end{bmatrix},$$

$$\hat{C}_1 = \begin{bmatrix} -1.7640 & -0.8190 \end{bmatrix}, \qquad \hat{C}_2 = \begin{bmatrix} 4.5977 & -0.8190 \end{bmatrix}.$$

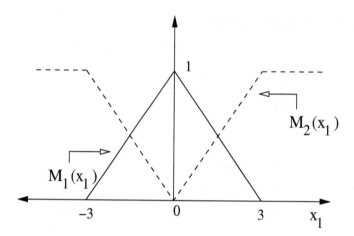

Figure 2. Membership functions for the two fuzzy set.

Hence, the resulting fuzzy filter is

$$E_\varepsilon \dot{\hat{x}}(t) = \sum_{i=1}^{2} \sum_{j=1}^{2} \mu_i \mu_j \hat{A}_{ij}(\varepsilon) \hat{x}(t) + \sum_{i=1}^{2} \mu_i \hat{B}_i y(t)$$

$$\hat{z}(t) = \sum_{i=1}^{2} \mu_i \hat{C}_i \hat{x}(t)$$

where

$$\mu_1 = M_1(x_1(t)) \quad \text{and} \quad \mu_2 = M_2(x_1(t)).$$

Case II: $v(t)$ are unavailable for feedback

In this case, $x_1(t) = v(t)$ is assumed to be unavailable for feedback; for instance, $J = [0 \; 1]$. This implies that μ_i is unavailable for feedback. Using the LMI optimization algorithm and Theorem 2 with $\varepsilon = 100 \; \mu H$, $\gamma = 0.6$ and $\delta = 1$, we obtain the following results:

$$\hat{A}_{11}(\varepsilon) = \begin{bmatrix} -2.3050 & -0.4186 \\ -32.3990 & -4.4443 \end{bmatrix}, \qquad \hat{A}_{12}(\varepsilon) = \begin{bmatrix} -2.3050 & -0.4186 \\ -32.3990 & -4.4443 \end{bmatrix},$$

$$\hat{A}_{21}(\varepsilon) = \begin{bmatrix} -2.3549 & -0.3748 \\ -32.4539 & -3.9044 \end{bmatrix}, \qquad \hat{A}_{22}(\varepsilon) = \begin{bmatrix} -2.3549 & -0.3748 \\ -32.4539 & -3.9044 \end{bmatrix},$$

$$\hat{B}_1 = \begin{bmatrix} -0.3053 \\ 3.9938 \end{bmatrix}, \qquad \hat{B}_2 = \begin{bmatrix} -0.3734 \\ 5.1443 \end{bmatrix},$$

$$\hat{C}_1 = \begin{bmatrix} 4.3913 & -0.1406 \end{bmatrix}, \qquad \hat{C}_2 = \begin{bmatrix} 1.9832 & -0.1406 \end{bmatrix}.$$

The resulting fuzzy filter is

$$E_\varepsilon \dot{\hat{x}}(t) = \sum_{i=1}^{2} \sum_{j=1}^{2} \hat{\mu}_i \hat{\mu}_j \hat{A}_{ij}(\varepsilon) \hat{x}(t) + \sum_{i=1}^{2} \hat{\mu}_i \hat{B}_i y(t)$$

$$\hat{z}(t) = \sum_{i=1}^{2} \hat{\mu}_i \hat{C}_i \hat{x}(t)$$

where

$$\hat{\mu}_1 = M_1(\hat{x}_1(t)) \quad \text{and} \quad \hat{\mu}_2 = M_2(\hat{x}_1(t)).$$

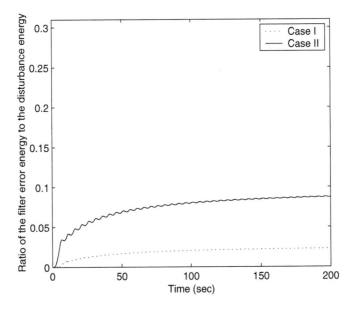

Figure 3. The ratio of the filter error energy to the disturbance noise energy:
$$\left(\frac{\int_0^{T_f} (z(t)-\hat{z}(t))^T (z(t)-\hat{z}(t)) dt}{\int_0^{T_f} w^T(t) w(t) dt} \right).$$

Remark 2. *The ratios of the filter error energy to the disturbance input noise energy are depicted in Figure 3 when* $\varepsilon = 100 \ \mu H$. *The disturbance input signal,* $w(t)$, *which was used during the simulation is the rectangular signal (magnitude 0.9 and frequency 0.5 Hz). Figures 4(a) - 4(b), respectively, show the responses of* $x_1(t)$ *and* $x_2(t)$ *in Cases I and II. Table I shows the performance index* γ *with different values of* ε *in Cases I and II. After 50 seconds, the ratio of the filter error energy to the disturbance input noise energy tends to a constant value which is about 0.02 in Case I and 0.08 in Case II. Thus, in Case I where* $\gamma = \sqrt{0.02} = 0.141$ *and in Case II where* $\gamma = \sqrt{0.08} = 0.283$, *both are less than the prescribed value 0.6. From Table 9.1, the maximum value of* ε *that guarantees the* $\mathcal{L}_2$-*gain of the mapping from the exogenous input noise to the filter error energy being less than 0.6 is 0.30 H, i.e.,* $\varepsilon \in (0, 0.30]$ *H in Case I, and 0.25 H, i.e.,* $\varepsilon \in (0, 0.25]$ *H in Case II.*   □

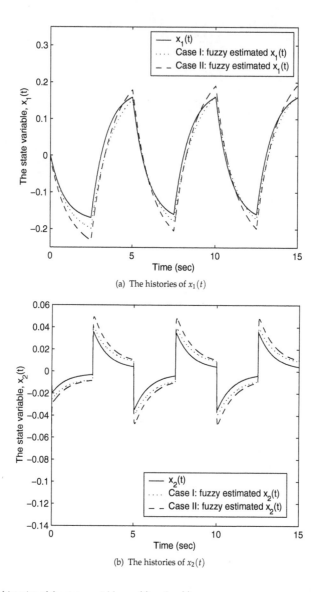

(a) The histories of $x_1(t)$

(b) The histories of $x_2(t)$

Figure 4. The histories of the state variables, $x_1(t)$ and $x_2(t)$.

	The performance index γ	
ε	Fuzzy Filter in Case I	Fuzzy Filter in Case II
0.0001	0.141	0.283
0.1	0.316	0.509
0.25	0.479	0.596
0.26	0.500	> 0.6
0.30	0.591	> 0.6
0.31	> 0.6	> 0.6

Table 1. The performance index γ of the system with different values of ε.

5. Conclusion

The problem of designing a robust $\mathcal{H}_\infty$ fuzzy ε-independent filter for a TS fuzzy descriptor system with parametric uncertainties has been considered. Sufficient conditions for the existence of the robust $\mathcal{H}_\infty$ fuzzy filter have been derived in terms of a family of ε-independent LMIs. A numerical simulation example has been also presented to illustrate the theory development.

Author details

Wudhichai Assawinchaichote
Department of Electronic and Telecommunication Engineering at King Mongkut's University of Technology Thonburi, Bangkok, Thailand

6. References

[1] H. K. Khalil, "Feedback control of nonstandard singularly perturbed systems," *IEEE Trans. Automat. Contr.*, vol. 34, pp. 1052–1060, 1989.

[2] Z. Gajic and M. Lim, "A new filtering method for linear singularly perturbed systems," *IEEE Trans. Automat. Contr.*, vol. 39, pp. 1952–1955, 1994.

[3] X. Shen and L. Deng, "Decomposition solution of $\mathcal{H}_\infty$ filter gain in singularly perturbed systems," *Signal Processing*, vol. 55, pp. 313–320, 1996.

[4] M.T. Lim and Z. Gajic, "Reduced-Order $\mathcal{H}_\infty$ optimal filtering for systems with slow and fast modes," *IEEE Trans. Circuits and Systems I*, vol. 47, pp. 250–254, 2000.

[5] P. Shi and V. Dragan, "Asymptotic $\mathcal{H}_\infty$ control of singularly perturbed system with parametric uncertainties," *IEEE Trans. Automat. Contr.*, vol. 44, pp. 1738–1742, 1999.

[6] P. V. Kokotovic, H. K. Khalil, and J. O'Reilly, *Singular Perturbation Methods in Control: Analysis and Design*, London: Academic Press, 1986.

[7] H. O. Wang, K. Tanaka, and M. F. Griffin, "An approach to fuzzy control of nonlinear systems: Stability and design issues," *IEEE Trans. Fuzzy Syst.*, vol. 4, pp. 14–23, 1996.

[8] K. Tanaka, T. Taniguchi, and H. O. Wang, "Fuzzy control based on quadratic performance function - A linear matrix inequality approach," in *Proc. IEEE Conf. Decision and Contr.*, pp. 2914–2919, 1998.

[9] B. S. Chen, C. S. Tseng, and H. J. Uang, "Mixed $\mathcal{H}_2/\mathcal{H}_\infty$ fuzzy output feedback control design for nonlinear dynamic systems: An LMI approach," *IEEE Trans. Fuzzy Syst.*, vol. 8, pp. 249–265, 2000.

[10] L. Xie, M. Fu, and C. E. de Souza, "$\mathcal{H}_\infty$ control and quadratic stabilisation of systems with parameter uncertainty via output feedback," *IEEE Trans. Automat. Contr.*, vol. 37, pp. 1253–1256, 1992.

[11] S. K. Nguang, "Robust nonlinear $\mathcal{H}_\infty$ output feedback control," *IEEE Trans Automat. Contr.*, vol. 41, pp. 1003–1008, 1996.

[12] S. K. Nguang and M. Fu, "Robust nonlinear $\mathcal{H}_\infty$ filtering," *Automatica*, vol. 32, pp. 1195–1199, 1996.

[13] K. Tanaka, T. Ikeda, and H. O. Wang, "Robust stabilization of a class of uncertain nonlinear systems via fuzzy control: Quadratic stabilizability, H_∞ control theory, and linear matrix inequality", *IEEE Trans. Fuzzy Syst.*, vol. 4, pp. 1–13, 1996.

[14] M. Teixeira and S. H. Zak, "Stabilizing controller design for uncertain nonlinear systems using fuzzy models", *IEEE Trans. Fuzzy Syst.*, vol. 7, pp. 133–142, 1999.

[15] S. H. Zak, "Stabilizing fuzzy system models using linear controllers", *IEEE Trans. Fuzzy Syst.*, vol. 7, pp. 236–240, 1999.

[16] L. X. Wang, *A course in fuzzy systems and control.* Englewood Cliffs, NJ: Prentice-Hall, 1997.

[17] S. K. Nguang and P. Shi, "$\mathcal{H}^\infty$ fuzzy output feedback control design for nonlinear systems: An LMI approach ," *IEEE Trans. Fuzzy Syst.*, vol. 11, pp. 331–340, 2003.

[18] S. K. Nguang and W. Assawinchaichote, "$\mathcal{H}_\infty$ filtering for fuzzy dynamic systems with pole placement," *IEEE Trans. Circuits Syst. I*, vol. 50, pp. 1503-1508, 2003.

[19] W.Assawinchaichote and S. K. Nguang, "$\mathcal{H}^\infty$ filtering for nonlinear singularly perturbed systems with pole placement constraints: An LMI approach", *IEEE Trans. Signal Processing*, vol. 52, pp. 579–588, 2004.

Chapter 6

Robust Stabilization for Uncertain Takagi-Sugeno Fuzzy Continuous Model with Time-Delay Based on Razumikhin Theorem

Yassine Manai and Mohamed Benrejeb

Additional information is available at the end of the chapter

1. Introduction

Fuzzy control systems have experienced a big growth of industrial applications in the recent decades, because of their reliability and effectiveness. Many researches are investigated on the Takagi-Sugeno models [1], [2] and [3] last decades. Two classes of Lyapunov functions are used to analysis these systems: quadratic Lyapunov functions and non-quadratic Lyapunov ones which are less conservative than first class. Many researches are investigated with non-quadratic Lyapunov functions [4]-[6], [7].

Recently, Takagi–Sugeno fuzzy model approach has been used to examine nonlinear systems with time-delay, and different methodologies have been proposed for analysis and synthesis of this type of systems [1]-[11], [12]-[13]. Time delay often occurs in many dynamical systems such as biological systems, chemical system, metallurgical processing system and network system. Their existences are frequently a cause of infeasibility and poor performances.

The stability approaches are divided into two classes in term of delay. The fist one tries to develop delay independent stability criteria. The second class depends on the delay size of the time delay, and it called delay dependent stability criteria. Generally, delay dependent class gives less conservative stability criteria than independent ones.

Two classes of Lyapunov-Razumikhin function are used to analysis these systems: quadratic Lyapunov-Razumikhin function and non-quadratic Lyapunov- Razumikhin ones. The use of first class brings much conservativeness in the stability test. In order to reduce the conservatism entailed in the previous results using quadratic function.

As the information about the time derivatives of membership function is considered by the PDC fuzzy controller, it allows the introduction of slack matrices to facilitate the stability analysis. The relationship between the membership function of the fuzzy model and the fuzzy controllers is used to introduce some slack matrix variables. The boundary information of the membership functions is brought to the stability condition and thus offers some relaxed stability conditions [5].

In this chapter, a new stability conditions for time-delay Takagi-Sugeno fuzzy systems by using fuzzy Lyapunov-Razumikhin function are presented. In addition, a new stabilization conditions for Takagi Sugeno time-delay uncertain fuzzy models based on the use of fuzzy Lyapunov function are presented. This criterion is expressed in terms of Linear Matrix Inequalities (LMIs) which can be efficiently solved by using various convex optimization algorithms [8],[9]. The presented methods are less conservative than existing results.

The organization of the chapter is as follows. In section 2, we present the system description and problem formulation and we give some preliminaries which are needed to derive results. Section 3 will be concerned to stability and stabilization analysis for T-S fuzzy systems with Parallel Distributed Controller (PDC). An observer approach design is derived to estimate state variables. Section 5 will be concerned to stabilization analysis for time-delay T-S fuzzy systems based on Razumikhin theorem. Next, a new robust stabilization condition for uncertain system with time delay is given in section 6. Illustrative examples are given in section 7 for a comparison of previous results to demonstrate the advantage of proposed method. Finally section 8 makes conclusion.

Notation: Throughout this chapter, a real symmetric matrix $S > 0$ denotes S being a positive definite matrix. The superscript "T" is used for the transpose of a matrix.

2. System description and preliminaries

Consider an uncertain T-S fuzzy continuous model with time-delay for a nonlinear system as follows:

$$IF \ z_1(t) \ is \ M_{i1} \ and...and \ z_p(t) \ is \ M_{ip}$$

$$THEN \quad \begin{cases} \dot{x}(t) = (A_i + \Delta A_i)x(t) + (D_i + \Delta D_i)x(t - \tau_i(t)) + (B_i + \Delta B_i)u(t) \\ x(t) = \phi(t), t \in [-\tau, 0] \end{cases} \quad (1)$$

where $M_{ij} (i = 1,2,...,r, j = 1,2,...,p)$ is the fuzzy set and r is the number of model rules; $x(t) \in \Re^n$ is the state vector, $u(t) \in \Re^m$ is the input vector, $A_i \in \Re^{n \times n}, D_i \in \Re^{n \times n}, B_i \in \Re^{n \times m}$, and $z_1(t),...,z_p(t)$ are known premise variables, $\phi(t)$ is a continuous vector-valued initial function on $[-\tau, 0]$; the time-delay $\tau(t)$ may be unknown but is assumed to be smooth function of time.. $\Delta A_i, \Delta D_i$ and ΔB_i are time-varying matrices representing parametric uncertainties in the plant model. These uncertainties are admissibly norm-bounded and structured.

$$0 \leq \tau(t) \leq \tau, \quad \dot{\tau}(t) \leq d \prec 1,$$

where $\tau \succ 0$ and d are two scalars.

The final outputs of the fuzzy systems are:

$$\dot{x}(t) = \sum_{i=1}^{r} h_i(z(t))\left\{(A_i + \Delta A_i)x(t) + (D_i + \Delta D_i)x(t - \tau_i(t)) + (B_i + \Delta B_i)u(t)\right\} \qquad (2)$$

$$x(t) = \phi(t), \quad t \in [-\tau, 0],$$

where

$$z(t) = [z_1(t)z_2(t)...z_p(t)]$$

$$h_i(z(t)) = w_i(z(t)) \bigg/ \sum_{i=1}^{r} w_i(z(t)), \; w_i(z(t)) = \prod_{j=1}^{p} M_{ij}(z_j(t)) \quad \text{for all } \underline{t}.$$

The term $M_{i1}(z_j(t))$ is the grade of membership of $z_j(t)$ in M_{i1}

Since
$$\begin{cases} \sum_{i=1}^{r} w_i(z(t)) > 0 \\ w_i(z(t)) \geq 0, \qquad i = 1,2,...,r \end{cases}$$

we have
$$\begin{cases} \sum_{i=1}^{r} h_i(z(t)) = 1 \\ h_i(z(t)) \geq 0, \qquad i = 1,2,...,r \end{cases} \qquad \text{for all } t.$$

The time derivative of premise membership functions is given by:

$$\dot{h}_i(z(t)) = \frac{\partial h_i}{\partial z(t)} \cdot \frac{\partial z(t)}{\partial x(t)} \cdot \frac{dx(t)}{dt} = \sum_{l=1}^{s} \upsilon_{il} \xi_{il} \times \frac{dx(t)}{dt}$$

We have the following property:

$$\sum_{k=1}^{r} \dot{h}_k(z(t)) = 0 \qquad (3)$$

Consider a PDC fuzzy controller based on the derivative membership function and given by the equation (4)

$$u(t) = -\sum_{i=1}^{r} h_i(z(t))F_i x(t) - \sum_{m=1}^{r} \dot{h}_m(z(t))K_m x(t) \qquad (4)$$

The fuzzy controller design consists to determine the local feedback gains F_i, and K_m in the consequent parts. The state variables are determined by an observer which detailed in next section.

By substituting (4) into (2), the closed-loop fuzzy system without time-delay can be represented as:

$$\dot{x}(t) = \sum_{i=1}^{r}\sum_{j=1}^{r} h_i\big(z(t)\big) h_j\big(z(t)\big)\left\{\left[A_{\Delta i} - B_{\Delta i}F_j - \sum_{m=1}^{r}\dot{h}_m\big(z(t)\big)B_{\Delta i}K_m\right]x(t) + D_{\Delta i}x\big(t-\tau_i(t)\big)\right\} \quad (5)$$

$$x(t) = \phi(t), \quad t \in \left[-\tau, 0\right],$$

where $A_{\Delta i} = A_i + \Delta A_i$; $D_{\Delta i} = D_i + \Delta D_i$ and $B_{\Delta i} = B_i + \Delta B_i$

The system without uncertainties is given by equation

$$\dot{x}(t) = \sum_{i=1}^{r}\sum_{j=1}^{r} h_i\big(z(t)\big) h_j\big(z(t)\big)\left\{\left[A_i - B_i F_j - \sum_{m=1}^{r}\dot{h}_m\big(z(t)\big)B_i K_m\right]x(t) + D_i x\big(t-\tau_i(t)\big)\right\} \quad (6)$$

$$x(t) = \phi(t), \quad t \in \left[-\tau, 0\right],$$

The open-loop system is given by the equation (7),

$$\dot{x}(t) = \sum_{i=1}^{r} h_i\big(z(t)\big)\big(A_{\Delta i}x(t) + D_{\Delta i}x\big(t-\tau_i(t)\big)\big) \quad (7)$$

$$x(t) = \phi(t), \quad t \in \left[-\tau, 0\right],$$

Assumption 1

The time derivative of the premises membership function is upper bounded such that $\left|\dot{h}_k\right| \leq \phi_k$, for $k = 1,\ldots,r$, where, $\phi_k, k = 1,\ldots,r$ are given positive constants.

Assumption 2

The matrices denote the uncertainties in the system and take the form of

$$\begin{cases} \Delta A_i = D_{a_i} F_{a_i}(t) E_{a_i} \\ \Delta B_i = D_{b_i} F_{b_i}(t) E_{b_i} \end{cases}$$

where $D_{a_i}, D_{b_i}, E_{a_i}$ and E_{b_i} are known constant matrices and $F_{a_i}(t)$ and $F_{b_i}(t)$ are unknown matrix functions satisfying :

$$\begin{cases} F_{a_i}^T(t) F_{a_i}(t) \leq I, \forall t \\ F_{b_i}^T(t) F_{b_i}(t) \leq I, \forall t \end{cases}$$

where I is an appropriately dimensioned identity matrix.

Lemma 1 (Boyd et al. Schur complement [16])

Given constant matrices Ω_1, Ω_2 and Ω_3 with appropriate dimensions, where $\Omega_1 = \Omega_1^T$ and $\Omega_2 = \Omega_2^T$, then

$$\Omega_1 + \Omega_3^T \Omega_2^{-1} \Omega_3 \prec 0$$

if and only if

$$\begin{bmatrix} \Omega_1 & \Omega_3^T \\ * & -\Omega_2 \end{bmatrix} \prec 0 \text{ or } \begin{bmatrix} -\Omega_2 & \Omega_3 \\ * & \Omega_1 \end{bmatrix} \prec 0$$

Lemma 2 (Peterson and Hollot [2])

Let $Q = Q^T, H, E$ and $F(t)$ satisfying $F^T(t) F(t) \leq I$ are appropriately dimensional matrices then the follow-ing inequality

$$Q + HF(t)E + E^T F^T(t) H^T \prec 0$$

is true, if and only if the following inequality holds for any $\lambda \succ 0$

$$Q + \lambda^{-1} HH^T + \lambda E^T E \prec 0$$

Theorem 1 (Razumikhin Theorem)[5]

Suppose $u, v, w : \Re^+ \to \Re^+$ are continuous, non-decreasing functions satisfying $u(s) \succ 0$, $v(s) \succ 0$ and $w(s) \succ 0$ for $s \succ 0$, $u(0) = v(0) = 0$, and v strictly increasing. If there exist a continuous function $V : \Re \times \Re^n \to \Re$ and a continuous non-decreasing function $p(s) \succ s$ for $s \succ 0$ such that

$$u(|x|) \leq V(t, x) \leq v(|x|), \qquad \forall t \in \Re, \ x \in \Re^n, \tag{8}$$

$$\dot{V}(t, x) \leq -w(|x|) \quad \text{if } V(t + \sigma, x(t + \sigma)) \leq p(V(t, x)), \quad \forall \sigma \in [-\tau, 0], \tag{9}$$

then the solution $x \equiv 0$ of (7) is uniformly asymptotically stable.

Lemma 3 [6]

Assume that $a \in \Re^{n_a}, b \in \Re^{n_b}, N \in \Re^{n_a \times n_b}$ are defined on the interval Ω. Then, for any matrices $X \in \Re^{n_a \times n_a}, Y \in \Re^{n_a \times n_b}$ and $Z \in \Re^{n_b \times n_b}$, the following holds:

$$-2 \int_\Omega a^T(\alpha) Nb(\alpha) d\alpha \leq \int_\Omega \begin{bmatrix} a(\alpha) \\ b(\alpha) \end{bmatrix}^T \begin{bmatrix} X & Y - N \\ Y^T - N^T & Z \end{bmatrix} \begin{bmatrix} a(\alpha) \\ b(\alpha) \end{bmatrix} d\alpha, \tag{10}$$

where $\begin{bmatrix} X & Y \\ Y^T & Z \end{bmatrix} \geq 0$.

Lemma 4 [9]

The unforced fuzzy time delay system described by **(7)** with u = 0 is uniformly asymptotically stable if there exist matrices $P \succ 0$, $S_i \succ 0$, X_{ai}, X_{di}, Z_{aij}, Z_{dij}, and Y_i, such that the following LMIs hold:

$$\begin{bmatrix} PA_i + A_i^T P + \tau\left(X_{ai} + X_{di}\right) + (2\tau + 1)P + Y_i + Y_i^T & -PD_i \\ Y_i^T - D_i^T P & -S_i \end{bmatrix} \prec 0 \qquad (11)$$

$$S_i \leq P \qquad (12)$$

$$A_j^T Z_{aij} A_j \leq P \qquad (13)$$

$$D_j^T Z_{dij} D_j \leq P \qquad (14)$$

$$\begin{bmatrix} X_{ai} & Y_i \\ Y_i^T & Z_{aij} \end{bmatrix} \geq 0 \qquad (15)$$

$$\begin{bmatrix} X_{di} & Y_i \\ Y_i^T & Z_{dij} \end{bmatrix} \geq 0 \qquad (16)$$

3. Basic stability and stabilization conditions

In order to design an observer for state variables, this section introduce two theorem developed for continuous TS fuzzy model for open-loop and closed-loop. First, consider the open-loop system without time-delay given by equation(17).

$$\dot{x}(t) = \sum_{i=1}^{r} h_i\left(z(t)\right) A_i x(t) \qquad (17)$$

The main approach for T-S fuzzy model stability is given in theorem follows. This approach is based on introduction of ε parameter which influences the stability region.

Theorem 2 [17]

Under assumption 1 and for $0 \leq \varepsilon \leq 1$, the Takagi Sugeno fuzzy system (17) is stable if there exist positive definite symmetric matrices $P_k, k = 1, 2, ..., r$, matrix $R = R^T$ such that the following LMIs hold.

$$P_k + R \succ 0, \quad k \in \{1,...,r\} \tag{18}$$

$$P_j + \mu R \succ 0, \quad j \in \{1,...,r\} \tag{19}$$

$$
\begin{aligned}
P_\phi + \frac{1}{2}\Big\{ A_i^T \left(P_j + \mu R \right) + \left(P_j + \mu R \right) A_i \\
+ A_j^T \left(P_i + \mu R \right) + \left(P_i + \mu R \right) A_j \Big\} \prec 0, \quad i \le j
\end{aligned}
\tag{20}
$$

where $i,j = 1,2,...,r$ and $P_\phi = \sum_{k=1}^{r} \phi_k \left(P_k + R \right)$ and $\mu = 1 - \varepsilon$

Proof

The proof of this theorem is given in detailed in article published in [17].

The closed-loop system without time delay is given by equation (21)

$$\dot{x}(t) = \sum_{i=1}^{r} h_i\big(z(t)\big) h_i\big(z(t)\big) G_{ii} x(t) + 2 \sum_{i=1}^{r} \sum_{i \prec j} h_i\big(z(t)\big) h_j\big(z(t)\big) \left\{ \frac{G_{ij} + G_{ji}}{2} \right\} x(t), \tag{21}$$

where

$$G_{ij} = A_i - B_i F_j \text{ and } G_{ii} = A_i - B_i F_i .$$

In this section we define a fuzzy Lyapunov function and then consider stability conditions.
A sufficient stability condition, for ensuring stability is given follows.

Theorem 2[18]

Under assumption 1, and assumption 2 and for given $0 \le \varepsilon \le 1$, the Takagi-Sugeno system
(21) is stable if there exist positive definite symmetric matrices $P_k, k = 1,2,...,r$, and R,
matrices $F_1,...,F_r$ such that the following LMIs hols.

$$P_k + R \succ 0, \quad k \in \{1,...,r\} \tag{22}$$

$$P_j + \mu R \ge 0, \quad j = 1,2,...,r \tag{23}$$

$$
P_\phi + \left\{ G_{ii}^T \left(P_k + \mu R \right) + \left(P_k + \mu R \right) G_{ii} \right\} \prec 0, \\
i,k \in \{1,...,r\}
\tag{24}
$$

$$
\left\{ \frac{G_{ij} + G_{ji}}{2} \right\}^T \left(P_k + \mu R \right) + \left(P_k + \mu R \right) \left\{ \frac{G_{ij} + G_{ji}}{2} \right\} \prec 0, \\
\text{for } i,j,k = 1,2,...,r \text{ such that } i \prec j
\tag{25}
$$

where

$$G_{ij} = A_i - B_i F_j, \quad G_{ii} = A_i - B_i F_i$$

And $P_\phi = \sum_{k=1}^{r} \phi_k (P_k + R)$

4. Observer design for T-S fuzzy continuous model

In order to determine state variables of system, this section gives a solution by the mean of fuzzy observer design.

A stabilizing observer-based controller can be formulated as follow:

$$\dot{\hat{x}}(t) = \sum_{j=1}^{r} h_i(z(t))\{A_i\hat{x}(t) + B_i u(t) + L_j(C_i\hat{x}(t) - y(t))\}$$

$$u(t) = \sum_{j=1}^{r} h_j(z(t))F_j\hat{x}(t)$$

(26)

The closed-loop fuzzy system can be represented as:

$$\dot{x}(t) = \sum_{i=1}^{r}\sum_{j=1}^{r} h_i(z(t))h_j(z(t))\left\{(A_i - B_i F_j) - \sum_{\rho=1}^{r} \dot{h}_\rho(z(t))(H_\rho + R)\right\}x(t)$$

$$+ \sum_{i=1}^{r}\sum_{j=1}^{r} h_i(z(t))h_j(z(t))\left\{B_i F_j + \sum_{\rho=1}^{r} \dot{h}_\rho(z(t))(H_\rho + R)\right\}e(t)$$

(27)

$$\dot{e}(t) = \sum_{i=1}^{r}\sum_{j=1}^{r} h_i(z(t))h_j(z(t))\{A_i - K_i C_j\}e(t)$$

(28)

The augmented system is represented as follows:

$$\dot{x}_a(t) = \sum_{i=1}^{r}\sum_{j=1}^{r} h_i(z(t))h_j(z(t))G_{ij}x_a(t)$$

$$= \sum_{j=1}^{r} h_i(z(t))h_j(z(t))G_{ii}x_a(t) + 2\sum_{i=1}^{r}\sum_{i<j}^{r} h_i(z(t))h_j(z(t))\left\{\frac{G_{ij} + G_{ji}}{2}\right\}x_a(t)$$

(29)

where

$$x_a(t) = \begin{bmatrix} x(t) \\ e(t) \end{bmatrix}$$

$$G_{ij} = \begin{bmatrix} A_i - B_i F_j - \sum_{\rho=1}^{r} \dot{h}_\rho B_i (H_\rho + R) & B_i F_j + \sum_{\rho=1}^{r} \dot{h}_\rho B_i (H_\rho + R) \\ 0 & A_i - K_i C_j \end{bmatrix}$$

By applying Theorem 2[18] in the augmented system (29) we derive the following Theorem.

Theorem 3

Under assumption 1 and for given $0 \le \mu \le 1$, the Takagi-Sugeno system (29) is stable if there exist positive definite symmetric matrices $P_k, k = 1, 2, \ldots, r$, and R, matrices $F_1, \ldots, F_r$ such that the following LMIs hols.

$$P_k + R \succ 0, \quad k \in \{1, \ldots, r\} \tag{30}$$

$$P_j + \mu R \ge 0, \quad j = 1, 2, \ldots, r \tag{31}$$

$$P_\phi + \left\{ G_{ii}^T \left(P_k + \mu R \right) + \left(P_k + \mu R \right) G_{ii} \right\} \prec 0, \\ i, k \in \{1, \ldots, r\} \tag{32}$$

$$\left\{ \frac{G_{ij} + G_{ji}}{2} \right\}^T \left(P_k + \mu R \right) + \left(P_k + \mu R \right) \left\{ \frac{G_{ij} + G_{ji}}{2} \right\} \prec 0, \\ \text{for } i, j, k = 1, 2, \ldots, r \text{ such that } i \prec j \tag{33}$$

where

$$G_{ij} = \begin{bmatrix} A_i - B_i F_j - \displaystyle\sum_{\rho=1}^{r} \dot{h}_\rho B_i \left(H_\rho + R \right) & B_i F_j + \displaystyle\sum_{\rho=1}^{r} \dot{h}_\rho B_i \left(H_\rho + R \right) \\ 0 & A_i - K_i C_j \end{bmatrix}$$

And $P_\phi = \displaystyle\sum_{k=1}^{r} \phi_k \left(P_k + R \right)$

Proof

The result follows immediately from the Theorem 2[18].

5. Stabilization of continuous T-S Fuzzy model with time-delay

The aim of this section is to prove the asymptotic stability of the time-delay system (6) based on the combination between Lyapunov theory and the Razumikhin theorem [5].

Theorem 4

Under assumption 1 and for given $0 \le \varepsilon \le 1$, the unforced fuzzy time delay system described by (7) with $u = 0$ is uniformly asymptotically stable if there exist matrices $P_k \succ 0, k = 1, 2, \ldots, r$, $S_i \succ 0$, X_{aij}, X_{di}, Z_{aij}, Z_{dij}, Y_i, and X, such that the following LMIs hold:

$$
\begin{bmatrix}
\begin{aligned}
& P_\beta + (P_k + \varepsilon X) G_{ij} + G_{ij}^T (P_k + \varepsilon X) \\
& + \tau (X_{aij} + X_{di}) + (2\tau + 1)(P_k + \varepsilon X) + Y_i + Y_i^T
\end{aligned}
& -(P_k + \varepsilon X) D_i \\[2ex]
Y_i^T - D_i^T (P_k + \varepsilon X) & -S_i
\end{bmatrix} \prec 0 \tag{34}
$$

where $\displaystyle P_\beta = \sum_{k=1}^{r} \beta_k (P_k + \varepsilon X)$.
$\quad\quad\quad G_{ij} = A_i - B_i F_j$

$$
S_i \le (P_k + \varepsilon X) \tag{35}
$$

$$
G_{ij}^T Z_{aij} G_{ij} \le (P_k + \varepsilon X) \tag{36}
$$

$$
D_j^T Z_{dij} D_j \le (P_k + \varepsilon X) \tag{37}
$$

$$
\begin{bmatrix} X_{aij} & Y_i \\ Y_i^T & Z_{aij} \end{bmatrix} \ge 0 \tag{38}
$$

$$
\begin{bmatrix} X_{di} & Y_i \\ Y_i^T & Z_{dij} \end{bmatrix} \ge 0 \tag{39}
$$

Proof

Let consider the fuzzy Lyapunov function as

$$
V(x) = x^T(t) V_k(x) x(t) \tag{40}
$$

$$
V_k(x) = \sum_{k=1}^{r} h_k (P_k + \varepsilon X)
$$

Given the matrix property, clearly,

$$
\lambda_{\min}(P_k + \varepsilon X)\|x(t)\|^2 \le x^T(t)(P_k + \varepsilon X)x(t) \le \lambda_{\max}(P_k + \varepsilon X)\|x(t)\|^2,
$$

where $\lambda_{\min(\max)}$ denotes the smallest (largest) eigenvalue of the matrix.

Finding the maximum value of $\sum_{k=0}^{r} h_k x^T(t)(P_k + \varepsilon X)x(t)$ is equivalent to determining the maximum value of $\sum_{k=0}^{r} h_k \lambda_{\max}(P_k + \varepsilon X)$.

Finding the minimum value of $\sum_{k=0}^{r} h_k x^T(t)(P_k + \varepsilon X)x(t)$ is equivalent to determining the minimum value of $\sum_{k=0}^{r} h_k \lambda_{\min}(P_k + \varepsilon X)$.

Define

$$\kappa_1 = \min_k \sum_{k=0}^{r} h_k \lambda_{\max}(P_k + \varepsilon X) \text{ for } 0 \le k \le r,$$

$$\kappa_2 = \max_k \sum_{k=0}^{r} h_k \lambda_{\min}(P_k + \varepsilon X) \text{ for } 0 \le k \le r.$$

Then,

$$\kappa_1 \|x(t)\|^2 \le \sum_{k=1}^{r} x^T(t)(P_k + \varepsilon X)x(t) \le \kappa_2 \|x(t)\|^2$$

In the following, we will prove the asymptotic stability of the time-delay system (7) based on the Razumikhin theorem [5].

Since

$$x(t) - x(t - \tau_i(t)) = \int_{t-\tau_i(t)}^{t} \dot{x}(s)ds,$$

The state equation of (7) with $u=0$ can be rewritten as

$$\dot{x}(t) = \sum_{i=1}^{r}\sum_{j=1}^{r} h_i h_j \left[(G_{ij} + D_i)x(t) - D_i \int_{t-\tau_i(t)}^{t} \dot{x}(s)ds \right]$$

where $G_{ij} = A_i - B_i F_j$

The derivative of V along the solutions of the unforced system (7) with $u = 0$ is thus given by

$$\dot{V} = x^T(t)\sum_{k=1}^{r} \dot{h}_k(P_k + \varepsilon X)x(t) + 2x^T(t)\sum_{i=1}^{r} h_i(P_i + \varepsilon X)\dot{x}(t) = \Upsilon_1(x,t) + \Upsilon_2(x,t) \quad (41)$$

$$\Upsilon_1(x,t) = x^T(t)\sum_{k=1}^{r} \dot{h}_k(P_k + \varepsilon X)x(t)$$

$$\Upsilon_2(x,t) = 2x^T(t)\sum_{k=1}^{r} h_k(P_k + \varepsilon X)\dot{x}(t) = 2x^T(t)\sum_{k=1}^{r} h_k(P_k + \varepsilon X) \times \sum_{i=1}^{r}\sum_{j=1}^{r} h_i h_j \left[(G_{ij} + D_i)x(t) - D_i \int_{t-\tau_i(t)}^{t} \dot{x}(s)ds \right].$$

Then, based on assumption 1, an upper bound of $\Upsilon_1(x,z)$ obtained as:

$$\Upsilon_1(x,z) \le \sum_{k=1}^{r} \beta_k \cdot x(t)^T (P_k + \varepsilon X)x(t) \quad (42)$$

and for $\Upsilon_2(x,t)$ we can written as,

$$\Upsilon_2(x,t) = 2\sum_{i=1}^{r}\sum_{j=1}^{r}h_ih_jx^T\sum_{k=1}^{r}h_k(P_k+\varepsilon X)(G_{ij}+D_i)x(t) - \sum_{i=1}^{r}\sum_{j=1}^{r}h_ih_j\int_{t-\tau_i(t)}^{t}\left\{2x^T(t)\sum_{k=1}^{r}h_k(P_k+\varepsilon X)D_i\right.$$

$$\left.\times\sum_{v=1}^{r}\sum_{\varsigma=1}^{r}h_v(s)h_\varsigma(s)\Big[G_{v\varsigma}x(s)+D_\varsigma x(s-\tau_j(s))\Big]ds\right\}$$

$$\Upsilon_2(x,t) = 2\sum_{i=1}^{r}\sum_{j=1}^{r}\sum_{k=1}^{r}h_ih_jh_kx^T\left\{(P_k+\varepsilon X)(G_{ij}+D_i)\right\}x(t) - \sum_{i=1}^{r}\sum_{j=1}^{r}\sum_{k=1}^{r}h_ih_jh_k\int_{t-\tau_i(t)}^{t}\left\{2x^T(t)(P_k+\varepsilon X)D_i\right.$$

$$\left.\times\sum_{v=1}^{r}\sum_{\varsigma=1}^{r}h_v(s)h_\varsigma(s)\Big[G_{v\varsigma}x(s)+D_\varsigma x(s-\tau_j(s))\Big]ds\right\} \tag{43}$$

Using the bounding method in(10), by setting $a=x(t)$ and $b=G_{ij}x(s)$, we have

$$-\int_{t-\tau_i(t)}^{t}2x^T(t)(P_k+\varepsilon X)D_i\times\sum_{v=1}^{r}\sum_{\varsigma=1}^{r}h_v(s)h_\varsigma(s)G_{v\varsigma}x(s)ds$$

$$\leq\tau_i(t)x^T(t)X_{ai}x(t)+2x^T(t)\big(Y_i-(P_k+\varepsilon X)D_i\big)\times\int_{t-\tau_i(t)}^{t}\sum_{v=1}^{r}\sum_{\varsigma=1}^{r}h_v(s)h_\varsigma(s)G_{v\varsigma}x(s)ds \tag{44}$$

$$+\int_{t-\tau_i(t)}^{t}\sum_{v=1}^{r}\sum_{\varsigma=1}^{r}h_v(s)h_\varsigma(s)x^T(s)G_{v\varsigma}^TZ_{aiv}G_{v\varsigma}x(s)ds$$

For any matrices X_{av}, Y_v and Z_{aiv} satisfying

$$\begin{bmatrix}X_{av} & Y_v \\ Y_v^T & Z_{aiv}\end{bmatrix}\geq 0$$

Similarly, it holds that

$$-\int_{t-\tau_i(t)}^{t}2x^T(t)(P_k+\varepsilon X)D_i\sum_{j=1}^{r}h_j(s)D_jx(s-\tau_j(s))ds$$

$$\leq\tau_i(t)x^T(t)X_{di}x(t)+2x^T(t)\big(Y_i-(P_k+\varepsilon X)D_i\big)\int_{t-\tau_i(t)}^{t}\sum_{j=1}^{r}h_j(s)D_jx(s-\tau_j(s))ds \tag{45}$$

$$+\int_{t-\tau_i(t)}^{t}\sum_{j=1}^{r}h_j(s)x^T(s-\tau_j(s))D_j^TZ_{dij}D_jx(s-\tau_j(s))ds$$

For any matrices X_{di}, Y_i and Z_{dij} satisfying

$$\begin{bmatrix} X_{di} & Y_i \\ Y_i^T & Z_{dij} \end{bmatrix} \geq 0$$

Hence, substituting (44) and (45) into (43), we have

$$\dot{V} \leq P_\beta + \sum_{i=1}^{r}\sum_{j=1}^{r}\sum_{k=1}^{r} h_i h_j h_k x^T(t)\Big[2\big(P_k + \varepsilon X\big)\big(G_{ij} + D_i\big) + \tau\big(X_{ai} + X_{di}\big)\Big]x(t)$$

$$+ \sum_{i=1}^{r}\sum_{k=1}^{r} h_i h_k 2x^T(t)\big(Y_i - \big(P_k + \varepsilon X\big)D_i\big) \times \int_{t-\tau_i(t)}^{t} \sum_{v=1}^{r}\sum_{\varsigma=1}^{r} h_v(s)h_\varsigma(s)\Big[G_{v\varsigma}x(s) + D_v x\big(s - \tau_v(s)\big)\Big]ds$$

$$+ \sum_{i=1}^{r}\sum_{k=1}^{r} h_i h_k \int_{t-\tau_i(t)}^{t} \sum_{v=1}^{r}\sum_{\varsigma=1}^{r} h_v(s)h_\varsigma(s)x^T(s)G_{v\varsigma}^T Z_{aiv} G_{v\varsigma}x(s)ds$$

$$+ \sum_{i=1}^{r}\sum_{k=1}^{r} h_i h_k \int_{t-\tau_i(t)}^{t} \sum_{j=1}^{r} h_j(s)x^T\big(s - \tau_j(s)\big)D_j^T Z_{dij} D_j x\big(s - \tau_j(s)\big)ds$$

$$\leq P_\beta + \sum_{i=1}^{r}\sum_{j=1}^{r} h_i h_j x^T(t)\Big[\big(P_k + \varepsilon X\big)G_{ij} + G_{ij}^T\big(P_k + \varepsilon X\big) + Y_i + Y_i^T + \tau\big(X_{aij} + X_{di}\big)\Big]x(t)$$

$$+ \sum_{i=1}^{r} h_i \Big[x^T(t)\big(Y_i - \big(P_k + \varepsilon X\big)D_i\big)S_i^{-1}\big(Y_i - \big(P_k + \varepsilon X\big)D_i\big)^T x(t) + x^T\big(t - \tau_i(t)\big)S_i x\big(t - \tau_i(t)\big)\Big]$$ (46)

$$+ \sum_{i=1}^{r} h_i \times \int_{t-\tau_i(t)}^{t} \sum_{v=1}^{r}\sum_{\varsigma=1}^{r} h_v(s)h_\varsigma(s)x^T(s)G_{v\varsigma}^T Z_{aiv} G_{v\varsigma}x(s)ds$$

$$+ \sum_{i=1}^{r} h_i \int_{t-\tau_i(t)}^{t} \sum_{j=1}^{r} h_j(s)x^T\big(s - \tau_j(s)\big)D_j^T Z_{dij} D_j x\big(s - \tau_j(s)\big)ds$$

Note that, by *Shur* complement, the LMI in (34) implies $L_i(\delta) \prec 0$ for a sufficiently small scalar $\delta \succ 0$, where

$$L_i(\delta) = P_\beta + \big(P_k + \varepsilon X\big)G_{ij} + G_{ij}^T\big(P_k + \varepsilon X\big) + Y_i + Y_i^T + \tau\big(X_{aij} + X_{di}\big)$$

$$+ \big(Y_i - \big(P_k + \varepsilon X\big)D_i\big)S_i^{-1}\big(Y_i - \big(P_k + \varepsilon X\big)D_i\big)^T x(t) + \big(2\tau + 1 + \tau\delta\big)\big(1 + \delta\big)\big(P_k + \varepsilon X\big)$$

In order to use the *Razumikhin* Theorem, suppose $V\big(x(t + \sigma)\big) \prec \big(1 + \delta\big)V\big(x(t)\big)$ for $\sigma \in [-\tau, 0]$. Then, if the LMIs in (35)–(39) also hold, we have from (46) that

$$\dot{V} \leq \sum_{i=1}^{r}\sum_{j=1}^{r} h_i h_j x^T(t)\Big[\big(P_k + \varepsilon X\big)G_{ij} + G_{ij}^T\big(P_k + \varepsilon X\big) + Y_i + Y_i^T + \tau\big(X_{aij} + X_{di}\big)\Big]x(t)$$

$$+ \sum_{i=1}^{r} h_i \Big[x^T(t)\big(Y_i - \big(P_k + \varepsilon X\big)D_i\big)S_i^{-1}\big(Y_i - \big(P_k + \varepsilon X\big)D_i\big)^T x(t) + x^T(t)\big(1 + \delta\big)\big(P_k + \varepsilon X\big)x(t)\Big]$$

$$+ \sum_{i=1}^{r} h_i \tau_i(t)x^T(t)\big(1 + \delta\big)\big(P_k + \varepsilon X\big)x(t) + \sum_{i=1}^{r} h_i \int_{t-\tau_i(t)}^{t} x^T(s)\big(1 + \delta\big)\big(P_k + \varepsilon X\big)x(s)ds$$

$$\leq \sum_{i=1}^{r} h_i x^T(t)\left[\left(P_k + \varepsilon X\right)A_i + A_i^T\left(P_k + \varepsilon X\right) + Y_i + Y_i^T + \tau\left(X_{ai} + X_{di}\right)\right]x(t)$$

$$+ \sum_{i=1}^{r} h_i\left[x^T(t)\left(Y_i - \left(P_k + \varepsilon X\right)D_i\right)S_i^{-1}\left(Y_i - \left(P_k + \varepsilon X\right)D_i\right)^T x(t) + x^T(t)(1+\delta)\left(P_k + \varepsilon X\right)x(t)\right]$$

$$+ \tau x^T(t)(1+\delta)\left(P_k + \varepsilon X\right)x(t) + \tau x^T(t)(1+\delta)^2\left(P_k + \varepsilon X\right)x(t)$$

$$= \sum_{i=1}^{r} h_i x^T(t) L_i(\delta) x(t)$$

$$\prec 0$$

which shows the motion of the unforced system **(7)** with $u = 0$ is uniformly asymptotically stable. This completes the proof.

6. Robust stability condition with PDC controller

Consider the closed-loop system (5). A sufficient robust stability condition for Time-delay system is given follow.

Theorem 5

Under assumption 1, and assumption 2 and for given $0 \leq \varepsilon \leq 1$, the Takagi-Sugeno system (5) is stable if there exist positive definite symmetric matrices $P_k, k = 1, 2, \ldots, r$, and R, matrices $F_1, \ldots, F_r$ such that the following LMIs hols.

$$P_k + R \succ 0, \quad k \in \{1, \ldots, r\} \tag{47}$$

$$P_j + \mu R \geq 0, \quad j = 1, 2, \ldots, r \tag{48}$$

$$\begin{bmatrix} \Phi_1 & \left(P_k + \mu R\right)D_{ai} & \left(P_k + \mu R\right)D_{bi} & \left(P_k + \mu R\right)\left(D_{di}\Delta_{di}E_{di}\right) \\ * & -\lambda I & 0 & 0 \\ * & * & -\lambda I & 0 \\ * & * & * & 0 \end{bmatrix} \prec 0 \tag{49}$$

$$i, k \in \{1, \ldots, r\}$$

with

$$\Phi_1 = P_\phi + \bar{G}_{ii}^T\left(P_k + \mu R\right) + \left(P_k + \mu R\right)\bar{G}_{ii} + \lambda\left(P_k + \mu R\right)\left[E_{ai}^T E_{ai} + \left(E_{bi}F_i\right)^T E_{bi}F_i\right]$$

$$
\begin{bmatrix}
\Phi_2 & & \\
* & -\lambda I & 0 \\
* & * & -\lambda I
\end{bmatrix} \prec 0
$$

$$
\begin{bmatrix}
\Phi_2 & (P_k + \mu R)(D_{ai} + D_{aj}) & (P_k + \mu R)(D_{bi} + D_{bj}) & (P_k + \mu R)(D_{di}\Delta_{di}E_{di}) \\
* & -\lambda I & 0 & 0 \\
* & * & -\lambda I & 0 \\
* & * & * & 0
\end{bmatrix} \prec 0 \tag{50}
$$

$$
i,k \in \{1,\ldots,r\}
$$

for $i,j,k = 1,2,\ldots,r$ such that $i \prec j$

with

$$
\Phi_2 = \left(\frac{\overline{G}_{ij} + \overline{G}_{ji}}{2}\right)^T (P_k + \mu R) + (P_k + \mu R)\left(\frac{\overline{G}_{ij} + \overline{G}_{ji}}{2}\right) + \lambda(P_k + \mu R)\left[(E_{ai} + E_{aj})^T (E_{ai} + E_{aj})\right.
$$

$$
\left. + (E_{bi}F_j + E_{bj}F_i)^T (E_{bi}F_j + E_{bj}F_i)^T\right]
$$

where $\overline{G}_{ij} = \left[A_i - B_iF_j - \sum_{m=1}^{r} \dot{h}_m(z(t))B_iK_m\right], G_{ii} = \left[A_i - B_iF_i - \sum_{m=1}^{r} \dot{h}_m(z(t))B_iK_m\right]$, $\mu = 1-\varepsilon$, and

$$
P_\phi = \sum_{k=1}^{r} \phi_k(P_k + R)
$$

Proof

Let consider the Lyapunov function in the following form:

$$
V(x(t)) = \sum_{k=1}^{r} h_k(z(t)) \cdot V_k(x(t)) \tag{51}
$$

with

$$
V_k(x(t)) = x^T(t)(P_k + \mu R)x(t), \quad k = 1,2,\ldots,r
$$

where

$$
P_k = P_k^T, R = R^T, 0 \le \varepsilon \le 1, \mu = 1-\varepsilon, \text{ and } (P_k + \mu R) \ge 0, \quad k = 1,2,\ldots,r.
$$

The time derivative of $V(x(t))$ with respect to t along the trajectory of the system (21) is given by:

$$
\dot{V}(x(t)) = \sum_{k=1}^{r} \dot{h}_k(z(t))V_k(x(t)) + \sum_{k=1}^{r} h_k(z(t))\dot{V}_k(x(t)) \tag{52}
$$

The equation (52) can be rewritten as,

$$\dot{V}\big(x(t)\big) = x^T(t)\left(\sum_{k=1}^{r}\dot{h}_k\big(z(t)\big)\big(P_k+\mu R\big)\right)x(t) + \dot{x}^T(t)\left(\sum_{k=1}^{r}h_k\big(z(t)\big)\big(P_k+\mu R\big)\right)x(t)$$
$$+ x^T(t)\left(\sum_{k=1}^{r}h_k\big(z(t)\big)\big(P_k+\mu R\big)\right)\dot{x}(t) \tag{53}$$

By substituting (5) into (53), we obtain,

$$\dot{V}\big(x(t)\big) = \Upsilon_1(x,z) + \Upsilon_2(x,z) + \Upsilon_3(x,z) \tag{54}$$

where

$$\Upsilon_1(x,z) = x^T(t)\left(\sum_{k=1}^{r}\dot{h}_k\big(z(t)\big)\cdot\big(P_k+\mu R\big)\right)x(t) \tag{55}$$

$$\Upsilon_2(x,z) = x^T(t)\sum_{k=1}^{r}\sum_{i=1}^{r}h_k\big(z(t)\big)h_i^2\big(z(t)\big)\times\left\{\overline{G}_{ii}^T\big(P_k+\mu R\big)+\big(P_k+\mu R\big)\overline{G}_{ii}\right\}x(t)$$
$$+ x^T(t)\sum_{k=1}^{r}\sum_{i=1}^{r}h_k\big(z(t)\big)h_i^2\big(z(t)\big)\times\left\{\left(\begin{bmatrix}D_{ai}&D_{bi}\end{bmatrix}\begin{bmatrix}\Delta_{ai}&0\\0&\Delta_{bi}\end{bmatrix}\begin{bmatrix}E_{ai}\\-E_{bi}F_i\end{bmatrix}\right)^T\big(P_k+\mu R\big)\right.$$
$$\left.+\big(P_k+\mu R\big)\left(\begin{bmatrix}D_{ai}&D_{bi}\end{bmatrix}\begin{bmatrix}\Delta_{ai}&0\\0&\Delta_{bi}\end{bmatrix}\begin{bmatrix}E_{ai}\\-E_{bi}F_i\end{bmatrix}\right)\right\}x(t) \tag{56}$$
$$+ x^T\big(t-\tau_i(t)\big)\sum_{k=1}^{r}\sum_{i=1}^{r}h_i\big(z(t)\big)h_k\big(z(t)\big)\left\{\big(D_{di}\Delta_{di}E_{di}\big)^T\big(P_k+\mu R\big)\right\}x(t)$$
$$+ x^T(t)\sum_{k=1}^{r}\sum_{i=1}^{r}h_i\big(z(t)\big)h_k\big(z(t)\big)\big(P_k+\mu R\big)\big(D_{di}\Delta_{di}E_{di}\big)x\big(t-\tau_i(t)\big)$$

$$\Upsilon_2(x,z) = \sum_{k=1}^{r}\sum_{i=1}^{r}h_k\big(z(t)\big)h_i^2\big(z(t)\big)\times\eta^T\Sigma_{ii}\eta$$

where $\eta^T = \begin{bmatrix}x^T(t) & x^T\big(t-\tau_i(t)\big)\end{bmatrix}$

$$\Sigma_{ii} = \begin{bmatrix}\Pi_1 & \big(P_k+\mu R\big)\big(D_{di}\Delta_{di}E_{di}\big)\\ \left\{\big(D_{di}\Delta_{di}E_{di}\big)^T\big(P_k+\mu R\big)\right\} & 0\end{bmatrix}$$

with $\Pi_1 = \left\{\overline{G}_{ii}^T\big(P_k+\mu R\big)+\big(P_k+\mu R\big)\overline{G}_{ii}\right\}$

$$+\left\{\left(\begin{bmatrix}D_{ai}&D_{bi}\end{bmatrix}\begin{bmatrix}\Delta_{ai}&0\\0&\Delta_{bi}\end{bmatrix}\begin{bmatrix}E_{ai}\\-E_{bi}F_i\end{bmatrix}\right)^T\big(P_k+\mu R\big)+\big(P_k+\mu R\big)\left(\begin{bmatrix}D_{ai}&D_{bi}\end{bmatrix}\begin{bmatrix}\Delta_{ai}&0\\0&\Delta_{bi}\end{bmatrix}\begin{bmatrix}E_{ai}\\-E_{bi}F_i\end{bmatrix}\right)\right\}x(t)$$

where $\overline{G}_{ii} = \begin{bmatrix}A_i - B_iF_i - \sum_{m=1}^{r}\dot{h}_m\big(z(t)\big)B_iK_m\end{bmatrix}$

$$\Upsilon_3(x,z) = x(t)^T \sum_{k=1}^{r}\sum_{i=1}^{r}\sum_{i<j} h_k(z(t))h_i(z(t))h_j(z(t)) \times \left\{ \left[\left(\frac{\bar{G}_{ij}+\bar{G}_{ji}}{2} \right)^T (P_k+\mu R) + (P_k+\mu R)\left(\frac{\bar{G}_{ij}+\bar{G}_{ji}}{2} \right) \right] \right\} x(t)$$

$$+ x(t)^T \sum_{k=1}^{r}\sum_{i=1}^{r}\sum_{i<j} h_k(z(t))h_i(z(t))h_j(z(t)) \times \left\{ \left[\begin{bmatrix} D_{ai} & D_{bi} \end{bmatrix} \begin{bmatrix} \Delta_{ai} & 0 \\ 0 & \Delta_{bi} \end{bmatrix} \begin{bmatrix} E_{ai} \\ -E_{bi}F_j \end{bmatrix} \right]^T (P_k+\mu R) \right.$$

$$+ (P_k+\mu R)\left[\begin{bmatrix} D_{ai} & D_{bi} \end{bmatrix} \begin{bmatrix} \Delta_{ai} & 0 \\ 0 & \Delta_{bi} \end{bmatrix} \begin{bmatrix} E_{ai} \\ -E_{bi}F_j \end{bmatrix} \right] \right\} x(t) + x(t)^T \sum_{k=1}^{r}\sum_{i=1}^{r}\sum_{i<j} h_k(z(t))h_i(z(t))h_j(z(t))$$

$$\times \left\{ \left[\begin{bmatrix} D_{aj} & D_{bj} \end{bmatrix} \begin{bmatrix} \Delta_{aj} & 0 \\ 0 & \Delta_{bj} \end{bmatrix} \begin{bmatrix} E_{aj} \\ -E_{bj}F_i \end{bmatrix} \right]^T (P_k+\mu R) + (P_k+\mu R)\left[\begin{bmatrix} D_{aj} & D_{bj} \end{bmatrix} \begin{bmatrix} \Delta_{aj} & 0 \\ 0 & \Delta_{bj} \end{bmatrix} \begin{bmatrix} E_{aj} \\ -E_{bj}F_i \end{bmatrix} \right] \right\} x(t) \quad (57)$$

$$+ x^T(t-\tau_i(t)) \sum_{k=1}^{r}\sum_{i=1}^{r} h_i(z(t))h_k(z(t)) \left\{ (D_{di}\Delta_{di}E_{di})^T (P_k+\mu R) \right\} x(t)$$

$$+ x^T(t) \sum_{k=1}^{r}\sum_{i=1}^{r} h_i(z(t))h_k(z(t))(P_k+\mu R)(D_{di}\Delta_{di}E_{di})x(t-\tau_i(t))$$

where $\bar{G}_{ij} = \left[A_i - B_i F_j - \sum_{m=1}^{r} \dot{h}_m(z(t))B_i K_m \right]$

$$\Upsilon_3(x,z) = \sum_{k=1}^{r}\sum_{i=1}^{r}\sum_{i<j} h_k(z(t))h_i(z(t))h_j(z(t)) \times \eta^T \Sigma_{ij} \eta$$

where $\eta^T = \begin{bmatrix} x^T(t) & x^T(t-\tau_i(t)) \end{bmatrix}$

$$\Sigma_{ij} = \begin{bmatrix} \Pi_2 & (P_k+\mu R)(D_{di}\Delta_{di}E_{di}) \\ \left\{ (D_{di}\Delta_{di}E_{di})^T (P_k+\mu R) \right\} & 0 \end{bmatrix}$$

with $\Pi_2 = \left\{ \left[\left(\frac{\bar{G}_{ij}+\bar{G}_{ji}}{2} \right)^T (P_k+\mu R) + (P_k+\mu R)\left(\frac{\bar{G}_{ij}+\bar{G}_{ji}}{2} \right) \right] \right.$

$$+ \left\{ \left[\begin{bmatrix} D_{ai} & D_{bi} \end{bmatrix} \begin{bmatrix} \Delta_{ai} & 0 \\ 0 & \Delta_{bi} \end{bmatrix} \begin{bmatrix} E_{ai} \\ -E_{bi}F_j \end{bmatrix} \right]^T (P_k+\mu R) + (P_k+\mu R)\left[\begin{bmatrix} D_{ai} & D_{bi} \end{bmatrix} \begin{bmatrix} \Delta_{ai} & 0 \\ 0 & \Delta_{bi} \end{bmatrix} \begin{bmatrix} E_{ai} \\ -E_{bi}F_j \end{bmatrix} \right] \right\}$$

$$+ \left\{ \left[\begin{bmatrix} D_{aj} & D_{bj} \end{bmatrix} \begin{bmatrix} \Delta_{aj} & 0 \\ 0 & \Delta_{bj} \end{bmatrix} \begin{bmatrix} E_{aj} \\ -E_{bj}F_i \end{bmatrix} \right]^T (P_k+\mu R) + (P_k+\mu R)\left[\begin{bmatrix} D_{aj} & D_{bj} \end{bmatrix} \begin{bmatrix} \Delta_{aj} & 0 \\ 0 & \Delta_{bj} \end{bmatrix} \begin{bmatrix} E_{aj} \\ -E_{bj}F_i \end{bmatrix} \right] \right\}$$

Then, based on assumption 1, an upper bound of $\Upsilon_1(x,z)$ obtained as:

$$\Upsilon_1(x,z) \le \sum_{k=1}^{r} \phi_k \cdot x(t)^T (P_k+\mu R)x(t) \quad (58)$$

Based on (3), it follows that $\sum_{k=1}^{r} \dot{h}_k(z(t))\varepsilon R = \bar{R} = 0$ where R is any symmetric matrix of proper dimension.

Adding $\bar{R}$ to (55), then

$$\Upsilon_1(x,z) \le \sum_{k=1}^{r} \phi_k \cdot x(t)^T (P_k + R) x(t) \tag{59}$$

Then,

$$\dot{V}(x(t)) \le \sum_{k=1}^{r} \phi_k x^T(t)(P_k + R)x(t) + \Upsilon_2(x,z) + \Upsilon_3(x,z)$$

If

$$\begin{bmatrix} H_{11} & (P_k + \mu R)D_{di}\Delta_{di}E_{di} \\ E_{di}^T\Delta_{di}^T D_{di}^T(P_k + \mu R) & 0 \end{bmatrix} \prec 0$$

where $H_{11} = \sum_{k=1}^{r} \phi_k (P_k + R) + \bar{G}_{ii}^T(P_k + \mu R) + (P_k + \mu R)\bar{G}_{ii}$

$$+ \left\{ \left(\begin{bmatrix} E_{ai} \\ -E_{bi}F_i \end{bmatrix} \right)^T \left(\begin{bmatrix} D_{ai} & D_{bi} \end{bmatrix} \right)^T (P_k + \mu R) + (P_k + \mu R) \begin{bmatrix} D_{ai} & D_{bi} \end{bmatrix} \begin{bmatrix} \Delta_{ai} & 0 \\ 0 & \Delta_{bi} \end{bmatrix} \begin{bmatrix} E_{ai} \\ -E_{bi}F_i \end{bmatrix} \right\}$$

Then, based on Lemma 2, an upper bound of H_{11} obtained as:

$$\sum_{k=1}^{r} \phi_k(P_k + R) + \bar{G}_{ii}^T(P_k + \mu R) + (P_k + \mu R)\bar{G}_{ii} + \lambda^{-1}(P_k + \mu R)\begin{bmatrix} D_{ai} & D_{bi} \end{bmatrix} \begin{bmatrix} D_{ai}^T \\ D_{bi}^T \end{bmatrix}$$

$$+ \lambda \begin{bmatrix} E_{ai}^T & -(E_{bi}F_i)^T \end{bmatrix} \begin{bmatrix} E_{ai} \\ -E_{bi}F_i \end{bmatrix} (P_k + \mu R) \prec 0$$

by Schur complement, we obtain,

$$\begin{bmatrix} \Phi_1 & (P_k + \mu R)D_{ai} & (P_k + \mu R)D_{bi} \\ * & -\lambda I & 0 \\ * & * & -\lambda I \end{bmatrix} \prec 0$$

with

$$\Phi_1 = P_\phi + \bar{G}_{ii}^T(P_k + \mu R) + (P_k + \mu R)\bar{G}_{ii} + \lambda(P_k + \mu R)\left[E_{ai}^T E_{ai} + (E_{bi}F_i)^T E_{bi}F_i \right]$$

$$\left\{ \left(\frac{\bar{G}_{ij} + \bar{G}_{ji}}{2} \right)^T (P_k + \mu R) + (P_k + \mu R)\left(\frac{\bar{G}_{ij} + \bar{G}_{ji}}{2} \right) \right\} + \left\{ \left(\begin{bmatrix} D_{ai} + D_{aj} & D_{bi} + D_{bj} \end{bmatrix} \begin{bmatrix} \Delta_{ai} + \Delta_{aj} & 0 \\ 0 & \Delta_{bi} + \Delta_{bj} \end{bmatrix} \right. \right.$$

$$\left. \times \begin{bmatrix} E_{ai} + E_{aj} \\ -E_{bi}F_j - E_{bj}F_i \end{bmatrix} \right)^T (P_k + \mu R) + (P_k + \mu R) \times \left(\begin{bmatrix} D_{ai} + D_{aj} & D_{bi} + D_{bj} \end{bmatrix} \begin{bmatrix} \Delta_{ai} + \Delta_{aj} & 0 \\ 0 & \Delta_{bi} + \Delta_{bj} \end{bmatrix} \times \begin{bmatrix} E_{ai} + E_{aj} \\ -E_{bi}F_j - E_{bj}F_i \end{bmatrix} \right) \right\}$$

$$\prec 0$$

Then, based on Lemma 2, an upper bound of $\Upsilon_1(x,z)$ obtained as:

$$
\left(\frac{\bar{G}_{ij}+\bar{G}_{ji}}{2}\right)^T (P_k+\mu R)+(P_k+\mu R)\left(\frac{\bar{G}_{ij}+\bar{G}_{ji}}{2}\right)+\lambda^{-1}(P_k+\mu R)\begin{bmatrix} D_{ai}+D_{aj} & D_{bi}+D_{bj} \end{bmatrix}\begin{bmatrix} D_{ai}^T+D_{aj}^T \\ D_{bi}^T+D_{bj}^T \end{bmatrix}
$$

$$
+\lambda\left[(E_{ai}+E_{aj})^T \quad (-E_{bi}F_j-E_{bj}F_i)^T\right]\times\begin{bmatrix} E_{ai}+E_{aj} \\ -E_{bi}F_j-E_{bj}F_i \end{bmatrix}(P_k+\mu R) \prec 0
$$

by Schur complement, we obtain,

$$
\begin{bmatrix} \Phi_2 & (P_k+\mu R)(D_{ai}+D_{aj}) & (P_k+\mu R)(D_{bi}+D_{bj}) \\ * & -\lambda I & 0 \\ * & * & -\lambda I \end{bmatrix} \prec 0
$$

with

$$
\Phi_2 = \left(\frac{\bar{G}_{ij}+\bar{G}_{ji}}{2}\right)^T (P_k+\mu R)+(P_k+\mu R)\left(\frac{\bar{G}_{ij}+\bar{G}_{ji}}{2}\right)+\lambda(P_k+\mu R)\left[(E_{ai}+E_{aj})^T(E_{ai}+E_{aj})\right.
$$

$$
\left.+(E_{bi}F_j+E_{bj}F_i)^T(E_{bi}F_j+E_{bj}F_i)^T\right]
$$

If (49) and (50) holds, the time derivative of the fuzzy Lyapunov function is negative.
Consequently, we have $\dot{V}(x(t)) \prec 0$ and the closed loop fuzzy system (5) is stable. This is
complete the proof.

7. Numerical examples

Consider the following T-S fuzzy system:

$$
\dot{x}(t)=\sum_{i=1}^{r} h_i(z(t))A_i x(t) \tag{60}
$$

with: $r=2$

the premise functions are given by:

$$
h_1(x_1(t))=\frac{1+\sin x_1(t)}{2}; \quad h_2(x_1(t))=\frac{1-\sin x_1(t)}{2}; \quad A_1=\begin{bmatrix} -5 & -4 \\ -1 & -2 \end{bmatrix}; \quad A_2=\begin{bmatrix} -2 & -4 \\ 20 & -2 \end{bmatrix};
$$

It is assumed that $\left|x_1(t)\right| \le \frac{\pi}{2}$. For $\xi_{11}=0, \xi_{12}=0.5, \xi_{21}=-0.5$, and $\xi_{22}=0$, we obtain

$$
P_1=\begin{bmatrix} 37.7864 & 26.8058 \\ 26.8058 & 36.2722 \end{bmatrix}; \quad P_2=\begin{bmatrix} 98.5559 & 28.7577 \\ 28.7577 & 22.9286 \end{bmatrix}; \quad R=\begin{bmatrix} -1.2760 & -2.2632 \\ -2.2632 & -0.6389 \end{bmatrix}
$$

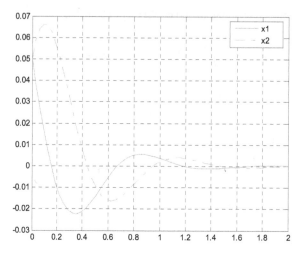

Figure 1. State variables

Figure 3 shows the evolution of the state variables. As can be seen, the conservatism reduction leads to very interesting results regarding fast convergence of this Takagi-Sugeno fuzzy system.

In order to show the improvements of proposed approaches over some existing results, in this section, we present a numerical example, which concern the feasibility of a time delay T-S fuzzy system. Indeed, we compare our fuzzy Lyapunov-Razumikhin approach (Theorem 3.1) with the Lemma 2.2 in [9].

Example 2. Consider the following T-S fuzzy system with u=0:

$$\dot{x}(t) = \sum_{i=1}^{2} h_i(z(t))\{A_i x(t) + D_i x(t - \tau_i(t))\}, \tag{61}$$

with:

$$A_1 = \begin{bmatrix} -2.1 & 0.1 \\ -0.2 & -0.9 \end{bmatrix}, \ A_2 = \begin{bmatrix} -1.9 & 0 \\ -0.2 & -1.1 \end{bmatrix}, \ D_1 = \begin{bmatrix} -1.1 & 0.1 \\ -0.8 & -0.9 \end{bmatrix}, \ D_2 = \begin{bmatrix} -0.9 & 0 \\ -1.1 & -1.2 \end{bmatrix},$$

with the following membership functions :

$$h_1 = \sin^2(x_1 + 0.5); \quad h_2 = \cos^2(x_1 + 0.5).$$

Assume that $\tau_i(t) = 0.5|\sin(x_1(t) + x_2(t) + 1)|$ where $x(t) = [x_1(t), x_2(t)]^T$. Then, $\tau_i(t) \leq \tau = 0.5$. Table 1. shows that our approach is less conservative than Lemma 2.2. given in [9].

Methods	τ_{max}
Lemma 2.1	0.6308
Theorem 3.1	$+\infty$

Table 1. Comparison results of maximum τ for Example 1

The LMIs in (34)-(39) are feasible by choosing $X_{ai} = X_a$, $X_{di} = X_d$, $Y_i = Y$, $Z_{aij} = Z_a$, $Z_{dij} = Z_d$, and $S_i = S$, $i, j = 1, 2$, and for $\tau = 0.5$ a feasible solution is given by

$$P_1 = \begin{bmatrix} 1.5121 & -0.1801 \\ -0.1801 & 1.1057 \end{bmatrix}, P_2 = \begin{bmatrix} 1.451 & -0.178 \\ -0.178 & 0.883 \end{bmatrix}, S = \begin{bmatrix} 1.021 & -0.064 \\ -0.064 & 0.664 \end{bmatrix},$$

$$Y = \begin{bmatrix} -0.611 & 0.169 \\ -0.243 & -0.421 \end{bmatrix}, X_a = \begin{bmatrix} 2.523 & 0.707 \\ 0.707 & 2.155 \end{bmatrix}, X_d = \begin{bmatrix} 1.448 & 0.094 \\ 0.094 & 2.353 \end{bmatrix}, Z_a = \begin{bmatrix} 0.201 & -0.087 \\ -0.087 & 0.369 \end{bmatrix},$$

$$Z_d = \begin{bmatrix} 0.849 & -0.227 \\ -0.227 & 0.246 \end{bmatrix},$$

8. Conclusion

This chapter provided new conditions for the stabilization with a PDC controller of Takagi-Sugeno fuzzy systems with time delay in terms of a combination of the Razumikhin theorem and the use of non-quadratic Lyapunov function as Fuzzy Lyapunov function. In addition, the time derivative of membership function is considered by the PDC fuzzy controller in order to facilitate the stability analysis. An approach to design an observer is derived in order to estimate variable states. In addition, a new condition of the stabilization of uncertain system is given in this chapter.

The stabilization condition proposed in this note is less conservative than some of those in the literature, which has been illustrated via examples.

Author details

Yassine Manai and Mohamed Benrejeb
National Engineering School of Tunis, LR-Automatique, Tunis, Tunisia

9. References

[1] T. Takagi, and M. Sugeno, "Fuzzy identification of systems and its application to modeling and control," *IEEE Trans. On System, Man and Cybernetics*, vol 15 (1), pp. 116–132, 1985.

[2] M.A.L. Thathachar, P. Viswanah, "On the Stability of Fuzzy Systems", *IEEE Transactions on Fuzzy Systems*, Vol. 5, N°1, pp. 145 – 151, February 1997.

[3] L. K. Wong, F.H.F. Leung, P.K.S. Tam, "Stability Design of TS Model Based Fuzzy Systems", *Proceedings of the Sixth IEEE International Conference on Fuzzy Systems*, Vol. 1, pp. 83–86, 1997.

[4] C.H. Fang, Y.S. Liu, S.W. Kau, L. Hong, and C.H. Lee, "A New LMI-Based Approach to Relaxed Quadratic Stabilization of T–S Fuzzy Control Systems," *IEEE Transactions on Fuzzy Systems*, Vol. 14, N° 3, pp.386–397, June 2006.

[5] H. K. Lam, F.H. Leung, "Stability Analysis of Fuzzy Model based Control Systems", Hong Kong, Springer 2011.

[6] I. Abdelmalek, N. Golea, and M.L. Hadjili, "A New Fuzzy Lyapunov Approach to Non–Quadratic Stabilization of Takagi-Sugeno Fuzzy Models," *Int. J. Appl. Math. Comput. Sci.*, Vol. 17, No. 1, 39–51, 2007.

[7] L.A. Mozelli, R.M. Palhares, F.O. Souza, and E.M. Mendes, "Reducing conservativeness in recent stability conditions of TS fuzzy systems," *Automatica*, Vol. 45, pp. 1580–1583, 2009.

[8] Y.Y. Cao and P.M. Frank, "Stability analysis and synthesis of nonlinear time-delay systems via Takagi–Sugeno fuzzy models", Fuzzy Sets and systems, Vol. 124 N°2, pp. 213-229, 2001.

[9] C. Lin, Q.G. Wang, T.H. Lee, "Delay-dependent LMI conditions for stability and stabilization of T–S fuzzy systems with bounded time-delay", Fuzzy Sets and Systems, Vol. 157 N°9, pp. 1229-1247, 2006.

[10] C. Lin, Q.G. Wang, and T. H. Lee, "Fuzzy Weighting-dependent approach to H_∞ filter design for Time-delay fuzzy systems", IEEE Transactions on Signal Processing, Vol. 55 N° 6, 2007.

[11] C. Lin, Q.G. Wang, and T. H. Lee, LMI Approach to Analysis and Control of Takagi–Sugeno Fuzzy Systems with Time Delay, Springer-Verlag, Berlin 2007.

[12] Y.Y. Cao, and P.M. Frank. "Analysis and synthesis of nonlinear time-delay systems via fuzzy control approach". IEEE Transactions on Fuzzy Systems, 8(2), 200-211, 2000.

[13] I Amri, D Soudani, M Benrejeb, "Exponential Stability and Stabilization of Linear Systems with Time Varying Delays",Conf. SSD'09, Djerba, 2009.

[14] K. Tanaka, T. Hori, and H.O. Wang, "A multiple Lyapunov function approach to stabilization of fuzzy control systems," IEEE Transactions on Fuzzy Systems, Vol. 11 N°4, pp. 582–589, 2003.

[15] C.H. Fang, Y.S. Liu, S.W. Kau, L. Hong, and C.H. Lee, "A New LMI-Based Approach to Relaxed Quadratic Stabilization of T–S Fuzzy Control Systems," IEEE Transactions on Fuzzy Systems, Vol. 14, N° 3, pp.386–397, June 2006.

[16] H.O. Wang, K. Tanaka, M. F. Griffin, "An Approach to Fuzzy Control of Nonlinear Systems: Stability and Design Issues", IEEE Transactions On Fuzzy Systems, Vol. 4, N°1, February 1996.

[17] Y. Manai, M. Benrejeb, "Stability for Continuous Takagi-Sugeno Fuzzy System based on Fuzzy Lyapunov Function', Conf. CCCA'11, Hammamet, 2011.

[18] M. Yassine, B. Mohamed, "New Condition of Stabilisation for Continuous Takagi-Sugeno Fuzzy System based on Fuzzy Lyapunov Function", International Journal of Control and Automation, Vol. 4 No. 3, September, 2011.

Chapter 7

Performance Evaluation of PI and Fuzzy Controlled Power Electronic Inverters for Power Quality Improvement

Georgios A. Tsengenes and Georgios A. Adamidis

Additional information is available at the end of the chapter

1. Introduction

In recent years, the increasing use of power electronics in the commercial and industry processes results in harmonics injection and lower power factor to the electric power system [Emanuel A E. (2004)]. Conventionally, in order to overcome these problems, passive R-L-C filters have been used. The use of this kind of filters has several disadvantages. Recently, due to the evolution in modern power electronics, new device called "shunt active power filter (SAPF)" was investigated and recognized as a viable alternative to the passive filters. The principle operation of the SAPF is the generation of the appropriate current harmonics required by the non-linear load.

For the reference currents generation, one of the best known and effective technique is the 'instantaneous reactive power theory' or 'p-q theory' [Czarnecki L S. (2006)]. In the literature, various modifications of p-q theory for the reference currents generation have been proposed [Salmeron P, Herrera R S, Vazquez J R. (2007)], [Kilic T, Milun S, et al. (2007)]. It is a common phenomenon in an electric power system, the grid voltages to be non-ideal [Segui-Chilet S S, Gimeno-Sales F J, et al. (2007)]. In such condition, the p-q theory is ineffective. To improve the efficiency of the p-q theory various reference currents generation techniques [Kale M, Ozdemir E. (2005)], [Tsengenes G, Adamidis G. (2011)] have been proposed. Except from the reference currents generation method, the current control method plays an important role to the overall system's performance. Plenty of methods have been used in the current control loop [Buso S, Malesani L, et al. (1998)] (e.g. ramp comparison, space vector modulation), one of which is the hysteresis current control. Hysteresis current controller compared to other current control methods has a lot of advantages such as robustness and simplicity [Tsengenes G, Adamidis G. (2010)].

The conventional reference currents generation techniques use a PI controller in order to regulate the dc bus voltage. The tuning of the PI controller requires precise linear mathematical model of the plant, which is very difficult to obtained, and it fails to perform satisfactorily under parameters variations, non-linearities, etc. To overcome these disadvantages, in recent years controllers which use artificial intelligent techniques have been implemented, like fuzzy logic controllers (FLCs) and artificial neural network (ANN) [Saad A, Zellouma L. (2009)], [El-Kholy E E, El-Sabbe, et al. (2006)], [Han ., Khan M M, Yao et al. (2010)], [Skretas S B, Papadopoulos D P (2009)]. The FLC surpasses the conventional PI controller due to its ability to handle non-imparities, its superior perform with a non-accurate mathematical model of the systems, and its robustness. In the literature some papers which implement FLCs in SAPF in order to improve the efficiency of the reference currents generation technique [Han Y, Khan M M, Yao et al. (2010)], [Jain S K, Agrawal P, et al. (2002)] and the current control loop have been reported [Mekri F, Machmoum M. (2010)], [Lin B R, Hoft R G. (1996)].

In this chapter a fuzzy logic controlled SAPF for current harmonics elimination is presented. The control scheme is based on two FLCs, the first one controls the dc bus voltage and the second one controls the output current of the inverter. Furthermore for the reference currents generation a modified version of the p-q theory is proposed, in order to improve the performance of the SAPF under non-ideal grid voltages. The performance of the proposed control scheme is evaluated through computer simulations using the software Matlab/Simulink under steady state and transient response. The superiority of the proposed fuzzy logic control scheme over the conventional control scheme is established both in steady state and transient response for current harmonics elimination and dc bus voltage of the SAPF respectively.

At the end a proposal for future investigation is presented. A combination between the fuzzy and the PI control is proposed. The new controller is called "fuzzy-tuned PI controller". The theoretical analysis and some simulation results are illustrated in order to verify the efficiency of the fuzzy-tuned PI controller.

2. Description of the proposed fuzzy control scheme

The main function of the SAPF is the current harmonics elimination and the reactive power compensation of the load. The general block diagram of a grid connected SAPF, as well as the detailed model of the control scheme is illustrated in Fig.1. The Reference currents generation method includes the dc bus voltage control which is the outer control loop. The current control method is the internal control loop which generates the appropriate switching pattern.

2.1. Reference currents generation method

For the reference currents generation a modified version of the p-q theory is used. One of the disadvantages of the p-q theory is the very poor efficiency of the method under non-ideal grid voltages. In this chapter the generation of three virtual grid voltages is proposed, one per phase as shown in figure 2. These virtual voltages will have the same amplitude as the fundamental harmonic (50 Hz) of the grid voltage, and will be synchronized with zero phase shifting compared with the corresponding grid voltages.

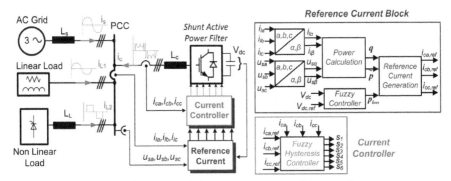

Figure 1. Synoptic diagram of the proposed electric power system and the control system

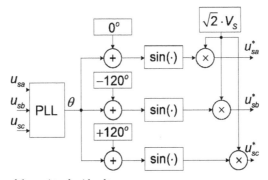

Figure 2. Generation of three virtual grid voltages

Mathematical equations for the virtual grid voltages a-b-c reference frame are given by equations (1), (2) and (3).

$$u_{sa}^* = \sqrt{2} \cdot V_s \cdot sin(\theta) \tag{1}$$

$$u_{sb}^* = \sqrt{2} \cdot V_s \cdot sin(\theta - 120^\circ) \tag{2}$$

$$u_{sc}^* = \sqrt{2} \cdot V_s \cdot sin(\theta + 120^\circ) \tag{3}$$

Where V_s is the root-mean-square (rms) value of the grid voltage ($V_s = \sqrt{u_{sa}^2 + u_{sb}^2 + u_{sc}^2}$), and θ is the angular frequency of the grid voltages ($\theta = 2 \cdot \pi \cdot f_{grid} = 2 \cdot \pi \cdot 50$).

The modified p-q theory, for the reference currents generation, will use the virtual grid voltages $u_{sa}^*, u_{sb}^*, u_{sc}^*$ and not the actual grid voltage. The load currents and the virtual grid voltages are transformed in α-β reference frame according to the transformer matrix of equation (4). The virtual grid voltages and the load current in α-β reference frame are given by equations (5), (6).

$$[C]_{abc \to \alpha\beta} = \sqrt{\frac{2}{3}} \cdot \begin{bmatrix} 1 & -\dfrac{1}{2} & \dfrac{1}{2} \\ 0 & \dfrac{\sqrt{3}}{2} & -\dfrac{\sqrt{3}}{2} \end{bmatrix} \tag{4}$$

$$\begin{bmatrix} u_{s\alpha}^* \\ u_{s\beta}^* \end{bmatrix} = [C]_{abc \to \alpha\beta} \cdot \begin{bmatrix} u_{sb}^* \\ u_{sb}^* \\ u_{sc}^* \end{bmatrix} \tag{5}$$

$$\begin{bmatrix} i_{l\alpha} \\ i_{l\beta} \end{bmatrix} = [C]_{abc \to \alpha\beta} \cdot \begin{bmatrix} i_{la} \\ i_{lb} \\ i_{lc} \end{bmatrix} \tag{6}$$

The instantaneous active and reactive powers of the electric power system are calculated via the following equation:

$$\begin{bmatrix} p \\ q \end{bmatrix} = \begin{bmatrix} u_{s\alpha}^* & u_{s\beta}^* \\ -u_{s\beta}^* & u_{s\alpha}^* \end{bmatrix} \cdot \begin{bmatrix} i_{l\alpha} \\ i_{l\beta} \end{bmatrix} \tag{7}$$

The instantaneous powers p and q are composed from a dc part ($\bar{\ }$) and an ac part ($\tilde{\ }$) corresponding to fundamental and harmonic current respectively. Equation (8) gives the instantaneous active and reactive power respectively.

$$\begin{aligned} p &= \tilde{p} + \bar{p} \text{ (a)} \\ q &= \tilde{q} + \bar{q} \text{ (b)} \end{aligned} \tag{8}$$

The ac component of the active power is extracted using a low pass filter. Using the p-q theory current harmonics are eliminated and the reactive power of the load is compensated. Therefore the reference currents of the SAPF in α-β reference frame are:

$$\begin{bmatrix} i_{c\alpha,ref} \\ i_{c\beta,ref} \end{bmatrix} = \frac{1}{(u_{s\alpha}^*)^2 + (u_{s\beta}^*)^2} \cdot \begin{bmatrix} u_{s\alpha}^* & -u_{s\beta}^* \\ u_{s\beta}^* & u_{s\alpha}^* \end{bmatrix} \begin{bmatrix} \tilde{p} - p_{loss} \\ q \end{bmatrix} \tag{9}$$

Where p_{loss} are related to the inverter operating losses. The grid should cover the p_{loss} in order to keep the capacitor voltage constant. Conventionally p_{loss} are calculated using a dc bus voltage sensor and a PI controller. In order to improve the dynamic performance of

SAPF and reduce the total harmonic distortion (THD$_i$) of the current a FLC for the dc bus
voltage control and a FLC for the current control are implemented. The current controller
handles the reference and the actual currents in a-b-c reference frame. As a result, the inverse
α-β transformation of equation (10) is used in order to transform the reference currents in a-
b-c reference frame.

$$
\begin{bmatrix} i_{ca,ref} \\ i_{cb,ref} \\ i_{cc,ref} \end{bmatrix} = \sqrt{\frac{2}{3}} \cdot \begin{bmatrix} 1 & 0 \\ -\frac{1}{2} & \frac{\sqrt{3}}{2} \\ -\frac{1}{2} & -\frac{\sqrt{3}}{2} \end{bmatrix} \cdot \begin{bmatrix} i_{\alpha,ref} \\ i_{\beta,ref} \end{bmatrix} \tag{10}
$$

2.2. Fuzzy logic dc bus voltage controller

For the dc bus voltage control a FLC is implemented. Figure 3 shows the synoptic block
diagram of the proposed FLC. As inputs to FLC the error between the sensed and the
reference dc bus voltage ($e = V_{dc,ref} - V_{dc}$) and the error variation ($\Delta e = e(k) - e(k-1)$) at
k^{th} sampling instant are used. The output of the fuzzy logic controller is considered as the
active power losses of the inverter (p_{loss}). The coefficients G$_1$, G$_2$ and G$_3$ are used to adjust
the input and output control signals.

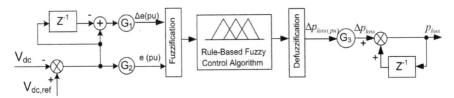

Figure 3. General structure of the fuzzy logic controller for dc bus voltage control

The FLC converts the crisp variables into linguistic variables. To implement this process it
uses the following seven fuzzy sets, which are: NL (Negative Large), NM (Negative
Medium), NS (Negative Small), Z (Zero), PS (Positive Small), PM (Positive Medium), PL
(Positive Large). The fuzzy logic controller characteristics used in this section are:

- Seven fuzzy sets for each input (e, Δe) and output (Δp$_{loss}$) with triangular and
  trapezoidal membership functions.
- Fuzzification using continuous universe of discourse.
- Implications using Mamdani's 'min' operator.
- Defuzzification using the 'centroid' method.

Figure 4 shows the normalized triangular and trapezoidal membership functions for the
input and output variables. The degree of fuzziness/membership ($\mu_{\delta,tri}(x)$) of the triangular
membership function is determined by equation (11.a). The degree of fuzziness ($\mu_{\delta,tra}(x)$) of
the trapezoidal membership function is determined by equation (11.b).

$$\mu_{\delta,tri}(x) = \begin{cases} 0 & x < a, x > c \\ \dfrac{x-a}{b-a} & a \le x < b \quad \text{(a)} \\ \dfrac{c-x}{c-b} & b \le x \le c \end{cases}$$

$$\mu_{\delta,tra}(x) = \begin{cases} 0 & x < a, x > d \\ \dfrac{x-a}{b-a} & a \le x \le b \\ 1 & b < x < c \quad \text{(b)} \\ \dfrac{d-x}{d-c} & c \le x \le d \end{cases} \qquad (11)$$

Where x, a, b, c, and d belong to the universe of discourse (X).

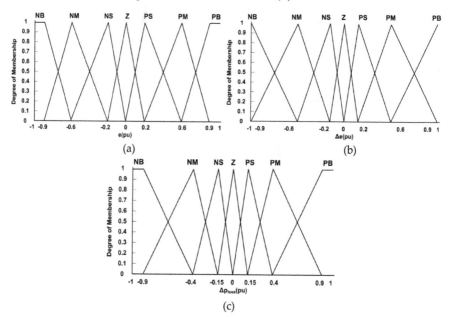

Figure 4. Membership functions for a) input variable e (pu), b) input variable Δe (pu), and c) output variable Δp_{loss} (pu)

Let $\mu_A(x)$ and $\mu_B(x)$ denote the degree of membership of the membership functions $\mu_A(\cdot)$ and $\mu_B(\cdot)$ of the input fuzzy sets A and B, where $x \in X$. Mamdani's logic operator is described as:

$$\phi\left[\mu_A(x), \mu_B(x)\right] = min\left[\mu_A(x), \mu_B(x)\right] = \mu_A(x) \wedge \mu_B(x) \qquad (12)$$

If $\mu_C(x)$ denotes the degree of membership of the membership functions $\mu_C(\cdot)$ of the output fuzzy sets C, where $x \in X$, equation 13 is used.

$$\mu_C(x) = \phi\left[\mu_A(x), \mu_B(x)\right] \cdot \mu_C(\cdot) \tag{13}$$

In the defuzzification procedure the centroid method with a discretized universe of discurse can be expressed as:

$$x_{out} = \frac{\displaystyle\sum_{i=1}^{n} x_i \cdot \mu_{out}(x_i)}{\displaystyle\sum_{i=1}^{n} \mu_{out}(x_i)} \tag{14}$$

Where x_{out} is crisp output value x_i is the output crisp variable and $\mu_{out}(x_i)$ is the degree of membership of the output fuzzy value, and i is the number of output discrete elements in the universe of discourse.

In the design of the fuzzy control algorithm, the knowledge of the systems behavior is very important. This knowledge is put in the form of rules of inference. The rule table which is shown in Table 1 contains 49 rules. The elements of the rule table are obtained from an understanding of the SAPF behavior [Jain S K, Agrawal P, et al. (2002)].

e Δe	NB	NM	NS	Z	PS	PM	PB
NB	NB	NB	NB	NB	NM	NS	Z
NM	NB	NB	NB	NM	NS	Z	PS
NS	NB	NB	NM	NS	Z	PS	PM
Z	NB	NM	NS	Z	PS	PM	PB
PS	NM	NS	Z	PS	PM	PB	PB
PM	NS	Z	PS	PM	PB	PB	PB
PB	Z	PS	PM	PB	PB	PB	PB

Table 1. Fuzzy control rule table.

2.3. Fuzzy logic Hysteresis current controller

One of the best known and most effective current control methods is the hysteresis band control technique. Some of its advantages are the simplicity of the construction combined with the excellent dynamic response. Apart from the significant advantages, this method has some drawbacks such as the high THD$_i$ index.

For the reduction of the THD$_i$ index, the implementation of a fuzzy logic hysteresis current controller is proposed. The synoptic diagram of fuzzy logic hysteresis controller for the phase-a is shown in figure 5. The same controller is applied to the other two-phases (b and c). As inputs to FLC the error between the reference current and the sensed current ($e = i_{ca,ref} - i_{ca}$) and the error variation ($\Delta e = e(k) - e(k-1)$) at k^{th} sampling instant are used. The output of the FLC is considered as the amplitude of the current error. The coefficients F_1 and F_2 are used to adjust the input control signals. The saturations blocks are used for limiting the initial error.

The fuzzy logic controller characteristics used in this section are the same as in the previous section (2.2). Figure 6 shows the triangular and trapezoidal membership functions for the input and output variables. The rule table for the hysteresis fuzzy logic controller is the same with Table 1.

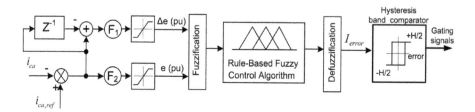

Figure 5. General structure of the hysteresis FLC for the current control loop

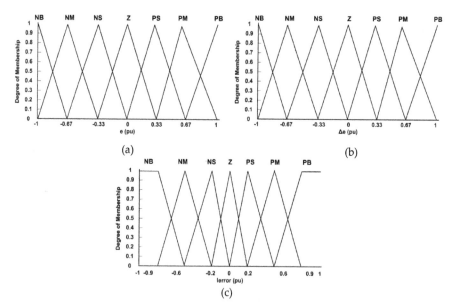

Figure 6. Membership functions for a) input variable e, b) input variable Δe, and b) output variable I_error

3. Non-ideal grid voltages

It is a common phenomenon in electric power system the grid voltages to be non-ideal. This problem is particularly important in Greek electric power distribution system, mainly due to the large increase of power electronic devices. In this section the mathematical model of the grid voltages under several non-ideal cases will briefly be presented.

The ideal grid voltages have sinusoidal waveform and can be represented as:

$$\begin{bmatrix} u_{sa} \\ u_{sb} \\ u_{sc} \end{bmatrix} = \sqrt{2} \cdot V_s \cdot \begin{bmatrix} sin(\omega t) \\ sin(\omega t - 120^0) \\ sin(\omega t + -120^0) \end{bmatrix} \tag{15}$$

The ideal grid voltages have only the fundamental frequency component.

When the three-phase grid voltages are unbalanced (u_{su}), the grid voltages can be expressed as positive and negative sequence components as shown in equation (12).

$$\begin{bmatrix} u_{sua} \\ u_{sub} \\ u_{suc} \end{bmatrix} = \begin{bmatrix} u_{sua+} \\ u_{sub+} \\ u_{suc+} \end{bmatrix} + \begin{bmatrix} u_{sua-} \\ u_{sub-} \\ u_{suc-} \end{bmatrix} \tag{16}$$

Where u_{sua+}, u_{sub+}, and u_{suc+} are positive sequence components and u_{sua-}, u_{sub-}, and u_{suc-} are negative sequence components.

It is a very common phenomenon in electric power distribution systems, voltages having non-ideal waveforms, and different levels of harmonics. When the three-phase grid voltages are distorted (u_{sd}), the grid voltages have harmonics components. In this scenario the distorted grid voltage can be represented as:

$$\begin{bmatrix} u_{sda} \\ u_{sdb} \\ u_{sdc} \end{bmatrix} = \begin{bmatrix} u_{sda,f} \\ u_{sdb,f} \\ u_{sdc,f} \end{bmatrix} + \begin{bmatrix} u_{sda,h} \\ u_{sdb,h} \\ u_{sdc,h} \end{bmatrix} \tag{17}$$

Where, $u_{sdb,f}$, and $u_{sdc,f}$ are positive sequence components and , $u_{sdb,h}$, and $u_{sdc,h}$ are harmonics components of the grid voltages.

When the three-phase grid voltages are distorted and unbalanced (u_{du}), the grid voltages contain harmonic components and unbalances. For this case, the distorted and unbalanced three-phase grid voltages are expressed as:

$$\begin{bmatrix} u_{sdua} \\ u_{sdub} \\ u_{sduc} \end{bmatrix} = \begin{bmatrix} u_{sda,f} \\ u_{sdb,f} \\ u_{sdc,f} \end{bmatrix} + \begin{bmatrix} u_{sda,h} \\ u_{sdb,h} \\ u_{sdc,h} \end{bmatrix} + \begin{bmatrix} u_{sua-} \\ u_{sub-} \\ u_{suc-} \end{bmatrix} = \begin{bmatrix} u_{sua+} \\ u_{sub+} \\ u_{suc+} \end{bmatrix} + \begin{bmatrix} u_{sda,h} \\ u_{sdb,h} \\ u_{sdc,h} \end{bmatrix} + \begin{bmatrix} u_{sua-} \\ u_{sub-} \\ u_{suc-} \end{bmatrix} \tag{18}$$

4. Simulation results

In this section the electric power system of figure 1 will be simulated. The simulation will be carried out via Matlab/Simulink. The characteristics of the electric power system are shown in Table 2. Four practical scenarios were examined in which the grid voltages are ideal, unbalanced, distorted and distorted-unbalanced. For the worst case, where the grid voltages are distorted-unbalanced the performance of the electric power system will be analyzed using the conventional and the fuzzy logic control system. The behavior of the PI controller and the FLC will be compared based on the dc bus voltage control.

Besides, the behavior of the conventional hysteresis controller and the hysteresis FLC will be compared based on the inverter output current control. For the comparison of the performance between the conventional control methods and the control methods with fuzzy logic theory the THD_i index in steady state, and the oscillation of the dc bus voltage during the transient response will be considered. Thereafter, the non-linear load will be called "Load_1" and the linear load will be called "Load_2". In all cases the transient response occurs at the same time (t=0.4 sec). It was considered that time t=0.4 sec in the electric power system, additionally to the initial non-linear load (Load_1) a linear load (Load_2) is connected.

Grid voltage (rms)	Vs=230V	Non-linear load	R₁=4Ω	SAPF inductance	L_c=1mH
Grid inductance	Ls=0.1mH	Linear Load	L₂=1mH	dc side capacitor	Cdc=3mF
Firing angle	α=10°	Linear Load	R₂=2Ω	dc bus voltage	Vdc=1 kV
Non-linear load	L₁=1mH	Non-linear load side impedance			LL=1mH

Table 2. Parameters of the electric power system.

4.1. Distorted-Unbalanced grid voltages

In this case the grid voltages are considered to be distorted-unbalanced, and they are expressed as:

$$\begin{bmatrix} u_{sda} \\ u_{sdb} \\ u_{sdc} \end{bmatrix} = \sqrt{2} \cdot V_s \cdot \begin{bmatrix} sin(\omega t) \\ sin(\omega t - 120^0) \\ sin(\omega t + -120^0) \end{bmatrix} + 13 \cdot \begin{bmatrix} sin(\omega t) \\ sin(\omega t - 120^0) \\ sin(\omega t + 120^0) \end{bmatrix} + 23 \cdot \begin{bmatrix} sin(5\omega t) \\ sin(5\omega t - 120^0) \\ sin(5\omega t + 120^0) \end{bmatrix} + 8 \cdot \begin{bmatrix} sin(7\omega t) \\ sin(7\omega t - 120^0) \\ sin(7\omega t + 120^0) \end{bmatrix} \quad (19)$$

Figure 7 shows the distorted-unbalanced grid voltages. Figures 8 and 9 show the grid currents (i_{sa}, i_{sb}, i_{sc}) and the reactive power of the grid respectively, without the application of the SAPF. In Table 3 the THD_i index of the grid currents for the loads 'Load_1' and 'Load_1+Load_2' is denoted.

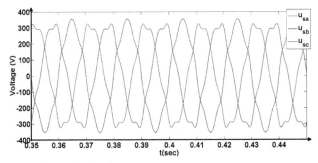

Figure 7. Distorted-unbalanced grid voltages

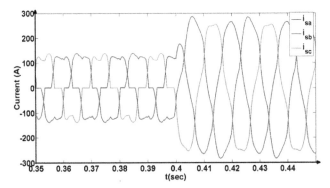

Figure 8. Grid currents without the application of the SAPF

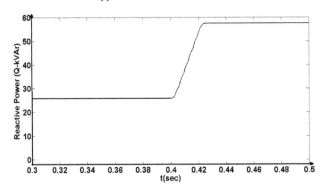

Figure 9. Reactive power of the grid without the SAPF

Phases	a	b	c
THD$_i$ (Load_1)	18.84	26.62	21.92
THD$_i$ (Load_1 + Load_2)	5.61	10.02	5.61

Table 3. Grid current THD$_i$ index.

4.1.1. Conventional controller and fuzzy controller for dc bus voltage control

In this section the control scheme consists of the PI dc bus voltage controller and the hysteresis current controller (called "PI-HYS") will be compared with the control scheme consists of the fuzzy dc bus voltage controller (from section 2.2) and the hysteresis current controller (called "FUZ-HYS"). Figure 10 shows the dc bus voltage transient response at time t=0.4 sec for both control schemes.

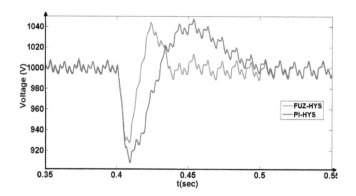

Figure 10. Dc bus voltage response for both control schemes

From figure 10 it is obvious that the dc bus voltage fuzzy logic controller outperforms the conventional PI controller. In particular, it is noted that the oscillation of the dc bus voltage with the application if the fuzzy logic controller is smaller compared to the PI one. Likewise the recovery time until the dc bus voltage returns to steady state is fairly smaller when the FLC is applied. This result has an effect on the time needed by the grid currents to return to steady state operation.

Figures 11 and 12 show the grid currents and the reactive power of the grid when the SAPF connected. Figures 11and 12 illustrate the results for both control schemes. Table 4 shows the THD$_i$ index of the grid currents considering the loads 'Load_1' and 'Load_1+Load_2'.

	PI-HYS			FYZ-HYS		
Phases	a	b	c	a	b	c
THD$_i$ (Load_1)	4.05	3.81	3.07	4.35	4.12	3.32
THD$_i$ (Load_1 + Load_2)	1.83	1.96	1.60	1.86	1.96	1.61

Table 4. Grid current THD$_i$ index.

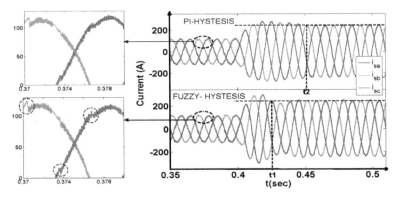

Figure 11. Grid currents with the application of the SAPF, for both control schemes

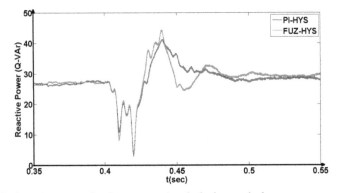

Figure 12. Grid reactive power after the compensation, for both control schemes

From figure 11 it is obvious that the faster response of the dc bus voltage using the fuzzy logic controller has a positive effect on the grid current, as the grid currents return to steady state operation faster (figure 11, time t_1 for FLC, and time t_2 for PI controller). It should be noted that, from figure 11 and Table 4 no significant change in the harmonic distortion of the grid currents in the case of dc bus voltage FLC is observed.

From the simulation results it is observed that the performance of the SAPF is satisfactory in the case where the grid voltages are distorted-unbalanced. This fact is a consequence of the modified version of the p-q theory, which was proposed in this chapter. The SAPF successfully eliminates the high order harmonics from the grid currents.

It is also observed that the SAPF compensates the reactive power of the load. As shown in figure 9 the reactive power of the grid without compensation for the 'Load_1' is Q=27 kVAr, then adding the 'Load_2' is increased to Q=58 kVAr. By using the active power filter, reactive power compensation is achieved for both initial and final load (the compensated reactive power of the grid is Q=28 VAr) as shown in figure 12. Comparing the two control schemes, PI-HYS and FYZ-HYS, similar behavior for the reactive power compensation is detected. For both control schemes there is a short transient period during the change of load.

4.1.2. Conventional dc bus voltage and ac current controller compared with fuzzy controller

For the reduction of the THDi index of the grid currents, authors propose a control scheme consists of a fuzzy logic dc bus voltage controller together with fuzzy logic hysteresis current controllers (as in section 2.3) (called "FUZ-FYZ HYS").

In this section the control scheme consists of the PI controller for the dc bus voltage control and the hysteresis controller for current control (PI-HYS) will be compared with the control scheme consists of the fuzzy controller for the dc bus voltage control and the fuzzy logic hysteresis controller for current control (FUZ-FUZ HYS). Figure 13 shows the dc bus voltage response at time t=0.4 sec for both control schemes.

From figure 13 it is obvious that the use of fuzzy logic for and ac output current control outperforms the control scheme of PI dc bus voltage control and hysteresis current control. In particular we observe that the oscillation of the dc bus voltage is smaller when the fuzzy logic dc bus voltage control and the fuzzy hysteresis current control is used. Likewise the interval time until the dc bus voltage returns to steady state operation is fairly smaller when the fuzzy logic scheme is applied. Comparing the results of figure 10 with those of figure 10, it is observed that the control schemes of FUZ- HYS and FUZ-FUZ HYS have no significant difference in the control of the dc bus voltage.

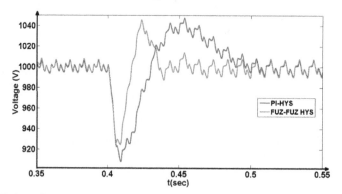

Figure 13. Dc bus voltage response, for both control schemes

Figures 14 and 15 show the grid currents and the reactive power of the grid with the application of the SAPF for both control schemes (PI-HYS and FUZ-FUZ HYS). Table 5 shows the grid currents THDi index for the loads 'Load_1' and 'Load_1+Load_2'.

	PI-HY			FYZ-HYS			FUZ-FYZ HYS		
Phases	a	b	c	a	b	c	a	b	c
THD$_i$ (1)	4.05	3.81	3.07	4.35	4.12	3.32	3.37	3.53	3.04
THD$_i$ (2)	1.83	1.96	1.60	1.86	1.96	1.61	1.62	1.77	1.57

Table 5. Grid current THD$_i$ index.

From figure 14 it is evident that the faster response of the dc bus voltage with the application of the FUZ-FUZ HYS control scheme has a positive effect on the grid current, as they return to steady state operation in smaller interval time (figure 14, time instant t$_1$ for FUZ-FUZ HYS control scheme, and time instant t$_2$ for PI-HYS control scheme). It should be noted that, from figure 14 smaller harmonic distortion of the grid currents is observed using the FUZ-FUZ HYS control scheme. In figure 14 some of the points where improvement is observed are highlighted using circles.

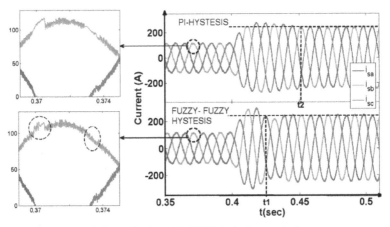

Figure 14. Grid currents with the application of the SAPF, for both control schemes

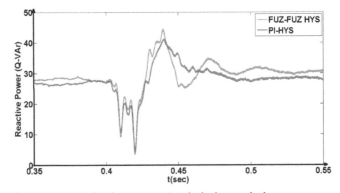

Figure 15. Grid reactive power after the compensation, for both control schemes

From the analysis of the simulation results, the improvement in the THDi index of the grid current using the FUZ-FUZ HYS control scheme is observed, as shown in Table 5. For phase-a, the improvement in the THDi index with the FUZ-FUZ HYS control scheme is about 22.5%. For phases-b and -c, the improvement in the THDi index with the FUZ-FUZ HYS control scheme is about 14.3% and 8.5% respectively. Investigating the electric power system, the unbalances in the THDi index is the result of the unbalances in the grid voltages.

From the comparison of the PI-HYS and FYZ-FUZ HYS control schemes, similar behavior for the reactive power compensation is observed. For both control schemes there is a short transition period during the load change. As shown in figure 9 the reactive power of the grid considering only the 'Load_1', without compensation for the is Q=27 kVAr, then adding the 'Load_2' the reactive power is increased to Q=58 kVAr. Using the SAPF, reactive power compensation is achieved for both initial and final load (in this case, the reactive power of the grid is approximately Q=29 VAr) as shown in figure 15.

From the simulation results it is observed that considering the above mentioned case the performance of the SAPF is excellent, as well as the performance of the SAPF is not affected by the distorted-unbalanced grid voltages. This fact is a consequence of the modified version of the *p-q* theory.

5. Future research

The fuzzy logic controller outperforms the conventional PI controller due to robustness and the superior transient response. However FLC have some significant disadvantages. The main drawback of the FLC is the requirement of an expert for the design of the membership functions and the fuzzy rules. To overcome this disadvantage, a novel artificial intelligent controller called "fuzzy-tuned PI controller" has been proposed in the literature of automation control [De Carli A, Linguori P, et al. (1994)], [Zhao Z-Y, Tomizuka M, et al. (1993)]. The fuzzy-tuned PI controller in figure 16 is a combination of the fuzzy controller and the PI controller. Using the fuzzy part we can estimate the gains K_p and K_p of the PI controller. Then the PI controller based on these gains outputs the reference signal. The fuzzy-tuned PI controller was initially applied for the speed control of the induction motor drives [Chen Y, Fu b, et al. (2008)] and the dc bus voltage control of the grid connected inverters [Suryanarayana H, Mishra MK (2008)].

No significant work, comparing the performance of the PI and fuzzy-tuned PI controller for the current control of a grid connected inverter, has been reported. In this section the PI and the fuzzy-tuned PI controller are applied to the inner current control loop. The criterion for the comparison of the two controllers are based on the transient response.

5.1. Fuzzy-tuned PI controller analysis for current control

The synoptic block diagram of the proposed fuzzy-tuned PI controller is illustrated in figure 16. As inputs to the fuzzy-tuned PI current controller are the actual and the reference currents. The current controller outputs the appropriate reference signal (reference voltage).

The reference voltage in α-β reference frame will be used by the Space Vector Modulation algorithm for the switching pattern generation.

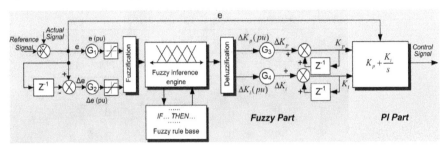

Figure 16. General structure of the fuzzy-tuned PI controller

The operation of the fuzzy-tuned PI controller is based on the use of a FLC for on-line tuning of the gains K_p and K_i of the PI controller, as shown in equation (20). Then the PI controller uses the adjusted gains K_p, K_i and the current error (e) to create the reference output control signals (referencevoltage).

$$K_p = \Delta K_p + K_p(k-1)$$
$$K_i = \Delta K_i + K_i(k-1)$$

(20)

As inputs to the fuzzy-tuned PI controller the error $e = i_{c,ref} - i_c$ and the error variation $\Delta e = e(k) - e(k-1)$ are determined. As outputs from the fuzzy part, the gains ΔK_i (pu) and ΔK_p (pu) of the PI controller, are determined. Using the gains K_i and K_i, the PI controller outputs the reference output voltage of the inverter ($u_{c,ref}$). The scaling factors G_1, G_2, G_3 and G_4 are used to normalize the input and output signals. In figure 17.a seven membership functions are used for each input (NL-Negative Large, NM-Negative Medium, NS-Negative Small, ZE-Zero, PS-Positive Small, PM-Positive Medium, and PL-Positive Large). In figure 17.b two membership functions are used for each output (B for Big and S for Small). For the fuzzy-tuned PI controller the triangular function was used as input and output fuzzy sets.

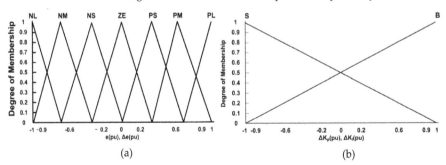

Figure 17. Membership functions for a) input variables e, Δe, and b) output variables ΔK_i (pu) and ΔK_p (pu)

The fuzzy rules which are shown in Table 6 for the ΔK_P and in Table 7 for the ΔK_i, are determined using the standard form of fuzzy rules: *IF e is A_i and Δe is B_j, THEN $\Delta K_{P,i}$ is C_{ij} and $\Delta K_{I,i}$ is D_{ij}.*

Δe	e						
	NL	NM	NS	ZE	PS	PM	PL
NL	B	B	B	B	B	B	B
NM	S	B	B	B	B	B	S
NS	S	S	B	B	B	S	S
ZE	S	S	S	B	S	S	S
PS	S	S	B	B	B	S	S
PM	S	B	B	B	B	B	S
PL	B	B	B	B	B	B	B

Table 6. Fuzzy control rules table for ΔK_P.

Δe	e						
	NL	NM	NS	ZE	PS	PM	PL
NL	B	B	B	B	B	B	B
NM	B	S	S	S	S	S	B
NS	B	B	S	S	S	B	B
ZE	B	B	B	S	B	B	B
PS	B	B	S	S	S	B	B
PM	B	S	S	S	S	S	B
PL	B	B	B	B	B	B	B

Table 7. Fuzzy control rules table for ΔK_I.

5.2. Comparison between fuzzy-tuned PI controller and the PI controller

In this section the behavior of the two current controllers will be compared based on the dynamic response. At the time instant t=0.4 sec a sudden variation of the output power of the inverter occurs. Figures 18.a and 18.b show the output currents of the inverter in *a-b-c* reference frame for the PI and the fuzzy-tuned PI current controller, respectively.

From figure 18 we can observe that the current error in *a-b-c* reference frame is very big
when the PI controller is used (the power of the inverter increases). When the fuzzy-tuned
PI controller is applied the error becomes almost zero while the recovery time is smaller
compared to the PI. This fact has a direct impact to the output currents, which in the case of
fuzzy-tuned PI controller have smoother behavior, while in the case of PI controller have
rougher behavior.

From the dynamic response of the electric power system is concluded that the fuzzy-tuned
PI controller is best suited for the inner current control loop.

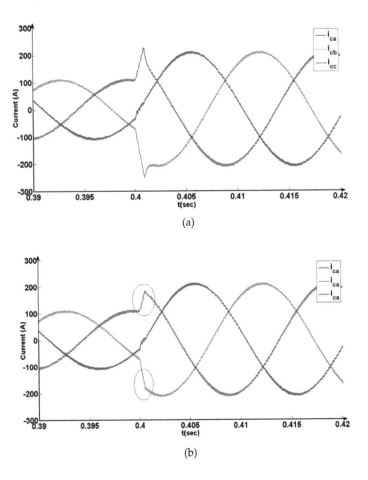

Figure 18. Current in ac side of the inverter during the dynamic response using the a) PI, and b) fuzzy-
tuned PI current controller

6. Conclusion

In this chapter a modified version of the p-q theory was proposed, in order to improve the performance of the SAPF in the case of non-ideal grid voltages. For the performance improvement of the control scheme, the fuzzy logic theory was applied. A fuzzy logic controller for the dc bus voltage control was used. From the computer simulations and the analysis of the results, smaller amplitude and duration of the dc bus voltage oscillations during the transient response has been demonstrated. A further investigation of the system was carried out applying fuzzy logic hysteresis controller to control the output current of the inverter. From the investigation of the control system using fuzzy logic controller both for dc bus voltage and inverter output current control, the dc bus voltage during the transient response and the THD$_i$ index of the grid currents are obviously improved.

Furthermore, authors use a combination of the PI and the fuzzy control known as "fuzzy-tuned PI" control. The performance of the inverter in case of the fuzzy-tuned PI control is used on the current control loop was investigated.

Author details

Georgios A. Tsengenes and Georgios A. Adamidis
Department of Electrical and Computer Engineering, Democritus University of Thrace, Greece

7. References

Buso S, Malesani L, Mattavelli P. (1998). Comparison of Current Control Techniques for Active Filter Applications. *IEEE Trans Ind Elect*, Vol.45, No.5, pp.722-729.

Chen Y, Fu b, Li Q (2008). Fuzzy logic based auto-modulation of parameters pi control for active power filter. In: World congress on intelligent control and automation, Chongqing, 2008.

Czarnecki L S. (2006). Instantaneous Reactive Power p-q Theory and Power Properties of Three-Phase System. *IEEE Trans Power Del*, Vol.21, No.1, pp.362-367.

De Carli A, Linguori P, Marroni A (1994). A fuzzy-pi control strategy. *Control Eng Practice*, Vol.2, No.1, pp.147-153.

El-Kholy E E, El-Sabbe, El-Hefnawy, Mharous M H. (2006). Three-phase active power filter based on current controlled voltage source inverter. Int J Electr Power Energy Syst, Vol.28, No.8 , pp.537-547.

Emanuel A E. (2004). Summary of IEEE Standard 1459: Definitions for the Measurement of Electric Power Quantities Under Sinusoidal, Nonsinusoidal, Balanced or Unbalanced Conditions. *IEEE Trans Ind Appl*, Vol.40, No.3, pp.869-876.

Han Y, Khan M M, Yao G, Zhou L-D, Chen C. (2010). A novel harmonic-free power factor corrector based on T-type APF with adaptive linear neural network (ADALINE) control. *Simulat Model Pract Theor*, Vol.16, No.9, pp.1215-1238.

Jain S K, Agrawal P, Gupta H O. (2002). Fuzzy logic controlled shunt active power filter for power quality improving. *IEE Electr Power Appl*, Vol.149, No.5, pp.317-328.

Kale M, Ozdemir E. (2005). Harmonic and reactive power compensation with shunt active power filter under non-ideal mains voltage. *Electr Power Syst Reseach*, Vol.74, No.3 pp.363-370.

Kilic T, Milun S, Petrovic G. (2007). Design and implementation of predictive filtering system for current reference generation of active power filter. Int *Journ of Electr Power Energy Syst*, Vol.29, No.2, pp.106-112.

Lin B R, Hoft R G. (1996). Analysis of power converter control using neural network and rule-based methods. *Electr Power Comp and Systems*, Vol.24, No.7, pp.695-720.

Mekri F, Machmoum M. (2010). A comparative study of voltage controllers for series active power filter. *Electr Power Syst Reseach*, Vol.80, No. , pp.615-626.

Saad A, Zellouma L. (2009). Fuzzy logic controller for three-level shunt active filter compensating harmonics and reactive power. *Electr Power Syst Reseach*, Vol. 79, No.10, pp.1337-1341.

Salmeron P, Herrera R S, Vazquez J R. (2007). A new approach for three-phase compensation based on the instantaneous reactive power theory. *Electr Power Syst Reseach*, Vol.78, No.4, pp.605-617.

Segui-Chilet S S, Gimeno-Sales F J, Orts S, Garcera G, Figueres E, Alcaniz M, Masot R. (2007). Approach to unbalance power active compensation under linear load unbalances and fundamental voltage asymmetries. *Int J Electr Power Energy Syst*, Vol.29, No.7, pp.526-539.

Singh G K, Singh A K, Mitra R. (2007). A simple fuzzy logic based robust active power filter for harmonics minimization under random load variation. *Electr Power Syst Reseach*, Vol.77, No.8 , pp.1101-1111.

Skretas S B, Dimitrios D. P. (2009). Efficient design and simulation of an expandable hybrid (wind–photovoltaic) power system with MPPT and inverter input voltage regulation features in compliance with electric grid requirements. *Electr Power Syst Reseach*, Vol.79, No.9 , pp.1271-1285.

Suryanarayana H, Mishra M K (2008). Fuzzy logic based supervision of dc link pi control in a dstatcom. *In: India conference*, Kanpur

Tsengenes G, Adamidis G. (2010). An Improved Current Control Technique for the Investigation of a Power System with a Shunt Active Filter. *In: International Symposium on Power Electronics, Electrical Drives, Automation and Motion*, Pisa, Italy.

Tsengenes G, Adamidis G. (2011). Investigation of the behavior of a three phase grid-connected photovoltaic system to control active and reactive power. *Elec Power Syst Reser*, Vol. 81, No. 1, pp.177-184, 2011.

Zhao Z-Y, Tomizuka M, Isaka S (1993). Fuzzy gain scheduling of pid controllers. *IEEE Trans Systems Man and Cybern*, Vol.23, No.4, pp. 1392-1398.

Zhou K, Wang D. (2002). Relationship between space-vector modulation and three-phase carrier-based PWM: A comprehensive analysis. *IEEE Trans Ind Electron*, Vol.49, No.1, pp.186-196.

Chapter 8

A Two-Layered Load and Frequency Controller of a Power System

Mavungu Masiala, Mohsen Ghribi and Azeddine Kaddouri

Additional information is available at the end of the chapter

1. Introduction

Automatic generation control (AGC) or called load frequency control (LFC) has gained a lot of interests in the past 30 decade (Benjamin &. Chan, 1982; Pan & Liaw, 1989; Kothari et al., 1989; Y. Wang et al., 1994; Indulkar & Raj, 1995; Karnavas & Papadopoulos, 2002; Moon et al., 2002; Sherbiny et al., 2003). LFC insures a sufficient and reliable supply of power with good quality. To ensure the quality of the power supply, it is necessary to deal with the control of the generator loads depending on the frequency with a proper LFC design. Therefore, the design of the controller is faced with nonlinear effects due to the physical components of the system, such as governor dead zone and generation rate constraints (GRC) and its complexity and the inherent characteristics of changing loads and parameters. Most actuators used in practice contain static (dead zone) or dynamic (backlash) non-smooth nonlinearities. These actuators are present in most mechanical and hydraulic systems such as servo valves. Their mathematical models are poorly known and limit the static and dynamic performance of feedback control system (Corradini & Orlando,2002). Conventional PI controller has been often used to achieve zero steady state frequency deviation. However, because of the load changing, the operating point of a power system may change very much during a daily cycle (Pan & Liaw, 1989). Therefore, a PI controller which is fixed and optimal when considering one operating point may no longer be suitable with various statuses. On the other hand, it is known that the classical LFC does not yield adequate control performance with consideration of the speed – governor non-smooth nonlinearities and GRC (Karnavas & Papadopoulos, 2002; Moon et al., 2002).

The problem of controlling systems with dead-zone nonlinearity has been addressed in the literature using various approaches some of which are dedicated to power systems. Reference (Tao & Kokotovic, 1994; X.-S. Wang et al., 2004) proposed adaptive schemes with and without dead zone inverse scheme, respectively, to track the error caused by the dead zone effect to zero. In the past decade, fuzzy logic controllers (FLC) have been developed successfully for analysis and control of nonlinear systems (Lee, 1990).

However, it has been sown by (Kim et al., 1994) that usual "Fuzzy PD" controller suffers from poor transient performance and a large steady state error when applied to systems with dead zones. On the other hand, when dealing with complex systems, the single-loop controller may not achieve the control performances and a multilayered controller turns out to be very helpful. The main advantage of the multilevel control lies in the freedom of the design of each layer (Yeh & Li, 2003; Oh & Park, 1998). The layers are designed to target particular objectives, so that design is simpler and performance improved. Motivated by the success of FLC, (Koo, 2001; Rubai, 1991) proposed new adaptive fuzzy controllers with online gain – tuning algorithm.

However, lots of computations are needed to calculate the adaptive control law with and without fuzzy system. To simplify the controller design (Kim et al., 1994) proposed a Two-layered fuzzy logic controller in which a fuzzy pre-compensator and a "fuzzy PD" controller were introduced to control plants with dead zones. Stimulated by ((Kim et al., 1994; L. X. Wang, 1997) designed a 2 layered fuzzy LFC (FLC-FLC) with the dead zone and GRC effects. Reference (Rubai, 1991) proposed a 2 layers fuzzy controller for the transient stability enhancement of the electric power system.

Based on the alternative choices proposed by (L. X. Wang, 1997) and the previous works (Kim et al., 1994; Sherbiny et al., 2003; Rubai, 1991], in this paper we study the case of a two-layer control architecture (FLC-CC) where the pre-compensator layer is constructed from fuzzy systems as a control supervisor and the other layer from the conventional method. In addition, we demonstrate that the proposed scheme exhibits a good transient and steady state performance, and is robust to load variations and system nonlinearities.

This paper is organized as follows. In section 2 we briefly introduce the systems investigated. Section 3 describes the idea underlying the approach and the design procedure of the proposed controller. The simulation plots that illustrate the behaviour of our scheme, taking into account parameters variations, GRC and speed governor dead zone are provided in section 4. Finally, conclusions based on extensive simulation results, recommendations and further research are drawn in section V.

2. Plant model

Power systems can be modeled by their power balance equations, linearized around the operating point. Since power systems are only exposed to small changes in load during their normal operation, a linear model can be used to design LFC. We consider the same single–area non reheat power system model as shown in Fig. 1. The investigated system consists of a speed–governor, a turbine that produces mechanical power, P_g, and the rotating mass (or power system). In steady state, P_g is balanced by the electrical power output, P_e, of the generator. Any imbalance between P_g and P_e produces accelerating power and thereby creates an incremental change in frequency, Δf. All parameters are given Appendix A.

The investigated model consists of a tandem-compound single non reheat turbine. The state space model can be expressed as following:

$$\dot{x}(t) = Ax(t) + BU(t) + F\Delta P_L \tag{1}$$

$$y(t) = Cx(t) \tag{2}$$

where:

$$A = \begin{bmatrix} -1/T_p & K_p/T_p & 0 & 0 \\ 0 & -1/T_t & 1/T_t & 0 \\ -1/RT_g & 0 & -1/T_g & -1/T_g \\ K_i & 0 & 0 & 0 \end{bmatrix} \tag{3}$$

$$B = \begin{bmatrix} 0 & 0 & 1/T_g & 0 \end{bmatrix}^T \tag{4}$$

$$F = \begin{bmatrix} -K_p/T_p & 0 & 0 & 0 \end{bmatrix}^T \tag{5}$$

$$C = [1\ 0\ 0\ 0] \tag{6}$$

The time constant, T_g, in the governor model is quite small and often it is neglected, which means that the governor is assumed to act very fast compared to the change in speed or frequency. This leads to a second order dynamic power system model. But for accuracy and comparison purposes, T_g is considered.

In linear control system theory, it is required that a state feedback controller:

$$\Delta Pc = -Kx(t) \tag{7}$$

Hence, the closed loop eigenvalues become insensitive to variations of the system parameters.

2.1. Model 1: Single-area non reheat power system

Consider the same isolated non reheat power system model reported in (Benjamin &. Chan, 1982; Pan & Liaw, 1989; Y. Wang et al., 1994) as shown in Figure 1, with the system parameters given in appendix A.

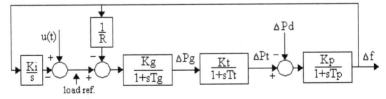

Figure 1. Block diagram of a isolated non reheat power system with supplementary control

Where u(t) denotes the existence of the proportional gain K_P, whereas its absence leads to an integral controller. The above described model has a tandem-compound single non reheat turbine and does not consider the speed-governor dead zone and GRC.

2.2. Model 2: two-areas reheat hydrothermal power system

The linearized mathematical model – 2 (see Fig.2), comprises an interconnection of two areas: single stages reheat thermal system (area 1) and a hydro system (area 2). The system parameters are given in appendix B. Figure 2 shows the small perturbation transfer function model of the hydro thermal system (Sherbiny et al., 2003). The speed governor dead zone and the GRC effects are also included in the model.

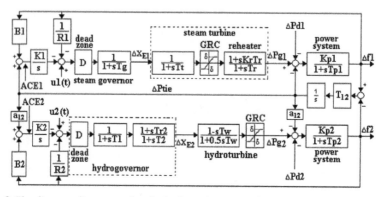

Figure 2. Bloc diagram of two-area reheat hydrothermal system with nonlinearities

3. Design of two layered controller for the system investigated

Considering the system shown in Figures 1, let P(s) represent the plant and D the speed governor actuator with dead zone (not present in Fig. 1). Recall that the supplementary PI control law can be written as following:

$$C_C[e(k), \int e(k)] = K_P e(k) + K_i \int e(k) \tag{8}$$

And in the case of "Fuzzy PD" controller, neglecting the scale factors, we get

$$C_F[e(k), \Delta e(k)] = F[e(k), \Delta e(k)] \tag{9}$$

C_C is a linear function of the error e(k) between the system output y_P(frequency deviation Δf) and the reference input Y_m (load reference ΔP_c) and the integral of the error whereas C_F is a function of the error and the change of error. From the above, we get

$$e(k) = y_m(k) - y_P(k) \quad \text{or} \quad y_P(k) = y_m(k) - e(k) \tag{10}$$

Let us assume that the supplementary control of the system is ensured by a "fuzzy PD" controller of the same type as (Indulkar & Raj, 1995; Karnavas & Papadopoulos, 2002).

3.1. Case 1: No actuator with dead zone

The plant output in this case can be written as

$$y_P(k) = P(s)\ F[e(k), \Delta e(k)] \tag{11}$$

Let $y_m(k) = y_m$. For steady state $\Delta e(k) = 0$ since the system is supposed to have reached the stabilizing time. Therefore, C_F becomes a function of $e(k)$ alone and (11) can be written as following:

$$y_P(k) = K_s\ F[e_{ss}, 0] = y_m - e_{ss} \tag{12}$$

where K_s is the system static gain and is given by $K_s = \lim_{s \to 0} P(s)$

By assuming that C_F is well tuned and that the load reference deviation $y_m = \Delta P_c = 0$, (12) becomes:

$$K_s\ F[e_{ss}, 0] = -e_{ss} \tag{13}$$

Taking into account the feedback negative input sign into the controller (see Fig. 1), it can be verified from the description of Fuzzy PD controller (Indulkar & Raj, 1995; Karnavas & Papadopoulos, 2002). that the law $F(\ .\ , 0)$ is an increasing odd function that can satisfy the following condition $f(x) = -f(-x)$, with $f(x) = F(\ .\ , 0)$. Therefore, it is clear that the solution to (13) is $e_{ss} = 0$, i.e., the steady state error of the system output is zero, as expected.

3.2. Case 2: Speed governor dead zone is present

The dead zone nonlinearity can be denoted as an operator is written as following:

$$u(k) = D(v(k)) \tag{14}$$

with $v(k)$ as input and $u(k)$ as output. The operator $D(v(k))$ has been described in detail by (Corradini & Orlando,2002; Tao & Kokotovic, 1994; X.-S. Wang et al., 2004; Oh & Park, 1998; Koo, 2001).The parameters of $D(v(k))$ are specified by the width 2d of the dead zone and the slop m of the response outside the dead zone. In this case, equation (12) can be written as :

$$y_P(k) = K_s\ D(v(k))\ F[e(k), 0] = y_m - e(k) \tag{15}$$

The solution to the equation (15) results in the steady state error as follow:

$$K_s\ D(v(k))\ F[e_{ss}, 0] - y_m = -e_{ss} \tag{16}$$

From (13) and (16) it can be seen that the steady state error e_{ss} in (16) is no longer zero. This is due to the presence of the dead zone in the speed governor actuator.

It has been demonstrated by (Yeh & Li, 2003) that the steady state error due to the dead zone in the actuator can be eliminated by adding some other constant η to the reference input y_m. We deliberately avoided using explicit knowledge of the value $D(v(k))$ because its parameters are poorly known or uncertain. Therefore, (16) becomes:

$$K_s D(v(k)) \ F[e + \eta \ , 0] - y_m = - e \tag{17}$$

Since PI controller is still the most used controller in power system (Pan & Liaw, 1989; Sherbiny et al., 2003) in our approach we use fuzzy logic rules to determine the appropriate value of η to be added to y_m. In this case, FLC plays the rule of a pre-compensator (supervisor) ensuring that the appropriate value of η is added to e(k) in order to eliminate the steady state error due to the dead zone. The conventional controller, called the stabilizer, present in the system, ensures the stabilization of the system.

The price to pay for changing the existing conventional controller into a FLC, which is proposed by (Sherbiny et al., 2003), would be in the computation. FLC is driven by a set of control rules rather than by two constant proportional and integral gains. As we shall see, the proposed scheme exhibits good transient and steady state behaviour. The proposed control scheme is depicted in Fig. 3 where C_1 is a FLC and C_2 a conventional controller. The feed-forward gain K_1 is normally set to the reciprocal of the K_s and constitutes an additional design parameter.

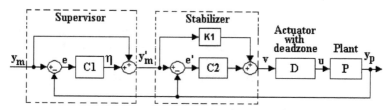

Figure 3. Proposed control structure

4. Design of the supervisor controller

As previously discussed, the first layer of the proposed control structure consists of the fuzzy logic based pre-compensator. The FLC law is based on standard fuzzy logic rules. It is well known that FLC consists of 3 stages, namely fuzzification, control rules inference engine and defuzzification. The reader is refereed to (Lee, 1990; L. X. Wang, 1997) for details on FLC. For LFC the process operator is assumed to respond to error e and change of error Δe (Indulkar & Raj, 1995); defined in (9). Considering the scale factors, (9) becomes as

$$C_F[e(k), \Delta e(k)] = F[n_e \ e(k), n_{\Delta e} \ \Delta e(k)] \tag{18}$$

where n_e and $n_{\Delta e}$ are the error and change of error scale factors respectively.

A label set corresponding to the linguistic variables control input e(k) and Δe(k) with a sampling time of 0.1 sec is as follows: L(e, Δe) = {NB, NM, ZE,PM, PB} where, NB - Negative Big, NM - Negative Medium, ZE – Zero, PM - Positive Medium and PB - Positive Big. The membership functions (MFs) for the control input variables are shown in Fig. 4. The universe of discourse of each control variable is normalised from -1 to 1. The proposed control structure uses the center of gravity defuzzification method to determine the output control as following:

$$C_F[e(k), \Delta e(k)] = n_\eta \; \frac{\sum\limits_{i}^{m} w_i y_i}{\sum\limits_{i}^{m} y_i} \tag{19}$$

Where n_η is the output control gain, w_i is the grade of the ith output MF, y_i is the output label for the value contributed by the ith MF, and m is the number of contributions from the rules. The fuzzy output variable is determined by same MFs shown in Fig. 4 and labelled as following $L(\eta) = \{NB, NM, ZE, PM, PB\}$

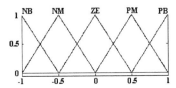

Figure 4. Membership functions of control input/output variables

The associated fuzzy matrices used in this work are given in table 1.

		e(k)				
		NB	NM	ZE	PM	PB
	NB	NB	NB	NB	NM	PM
	NM	NB	NB	NM	ZE	PM
	ZE	NB	NM	ZE	PM	PB
Δe(k)	PM	NM	ZE	PM	PB	PB
	PB	NM	PM	PB	PB	PB

Table 1. Fuzzy logic rules for pre-compensator C1

The performance of the FLC is affected by scaling factors of the inputs/output variables, MFs and the control rules. The selection of the optimum values of these factors is necessary in order to achieve satisfactory response [6]. But for the control system shown in Fig. 3, we can design the FLC without considering stability and use the stabilizer layer to deal with stability related problems.

5. Second layer: Stabilizer

The stabilizing layer consists of a conventional controller which is described in (8) and its design procedure is detailed in (Karnavas & Papadopoulos, 2002). The input to the present layer is the y'_m as shown in Fig. 3.

6. Simulation results

The power systems under investigation are simulated and subjected to different load disturbances in order to validate the effectiveness of the proposed scheme. The nonlinear effects, such as GRC and speed governor dead-zone, are also included in the simulations. The proposed controller (FLC-CC) will be evaluated qualitatively and quantitatively with

A single layer "PD FLC" (FLC) proposed in the first part of the work of (Karnavas & Papadopoulos, 2002) and a single layer conventional control (CC). The reader is referred to (Sherbiny et al., 2003) in order to compare our controller responses with the two-layered fuzzy controller proposed by (Sherbiny et al., 2003). The parameters of our controller for the two investigated models are given in table 2.

		n_e	$n_{\Delta e}$	n_η	K_p	K_i
Model 1		0.25	10.0	0.14	0.425	0.212
Model 2	area 1	10.0	20.0	0.15	4.00	6.00
	area 2	10.0	10.0	0.05	30.0	20.0

Table 2. Proposed controller parameters

6.1. Model 1 (Appendix A)

A step load perturbation of 10% (ΔP_d = 0.1 p.u.) of the nominal loading is considered. Fig 5 shows the simulation results of the frequency deviation Δf response to the step load change. The responses, obtained by the PD Fuzzy and the conventional controller (Karnavas & Papadopoulos, 2002) are also shown for comparison purpose.

Assume that the parameters R, Tg, Tt, Tp, Kp are subjected to a simultaneous changes of +30% from their nominal values. The frequency deviation response of the system is plotted in Fig. 6.

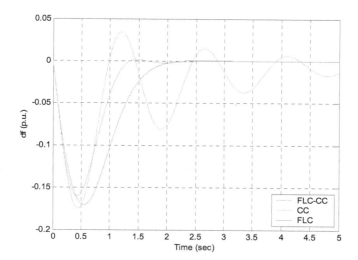

Figure 5. Frequency deviation for a load change of 0.1 p.u.

When GRC is applied to the system, its dynamic responses experience longer transient setting time t_s and larger overshoots OS compare to cases without the GRC. GRC of 3% p.u. MW/min and 10% p.u. MW/min are usually applied to reheat and non reheat turbines, respectively. In addition to GRC, the dead zone effect is also added to the system investigated. A dead zone width of 0.05 p.u. is considered. GRC and dead zone are taken into account by adding limiters to the turbines and an actuator to the speed governor input, respectively. Fig. 7 plots the responses of the system under nonlinear effects and a step load perturbation $\Delta P_d = 0.05$ p.u.

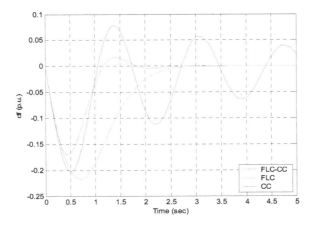

Figure 6. Frequency deviation at parameter changes of +30%

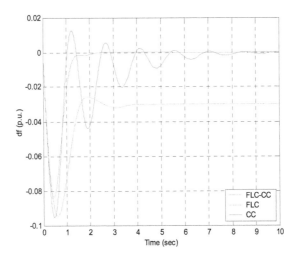

Figure 7. Frequency deviation due to $\Delta P_d = 0.05$ p.u with nonlinear effects

From Figs. 5-7 it can be observed that the proposed controller acts as fast as the FLC-FLC controller with less oscillatory, less undershoot and setting time. In addition the proposed scheme is also robust to load and parameters changes with and without nonlinearities. Fig. 7 shows the system steady state error for the PD FLC as previously predicted.

6.2. Model 2: Appendix (B)

Fig 8-10 show the simulation results of the frequency deviations and the tie line power responses of the two area power system due to a step load perturbation $\Delta P_d = 0.05$ p.u. without non-linear effects. The responses obtained by the conventional controller (CC) are also shown for comparison purpose.

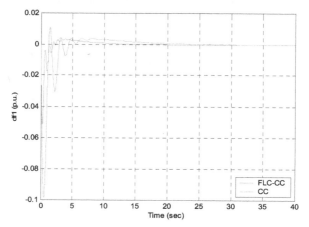

Figure 8. Frequency deviation of area 1 due to $\Delta P_{d1} = 0.05$ p.u.

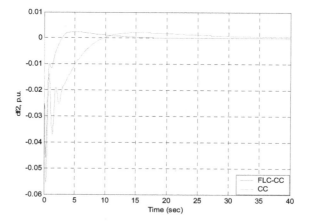

Figure 9. Frequency deviation of area 2 due to $\Delta P_{d2} = 0.05$ p.u

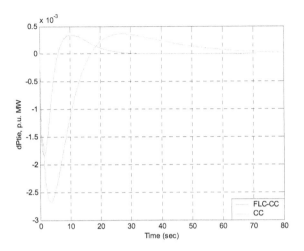

Figure 10. Tie line power deviation response due to ΔP_{d2}

It can be observed from the results obtained in Figs 8-10 that the proposed controller exhibits less oscillations and setting time compare to the conventional controller. Our controller responses are faster with smaller overshoot than the FLC-FLC controller. The reader is referred to (Sherbiny et al., 2003) for comparison.

Now assume that a dead zone width of 0.5 p.u. and GRC effects are considered in both area 1 and 2 simultaneously. Figs 11-13 plot the responses of the system under nonlinear effects.

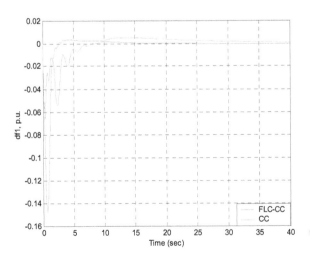

Figure 11. Frequency deviation of area 1 due to $\Delta P_{d1} = 0.05$ p.u. with nonlinear effects

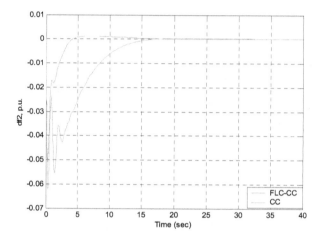

Figure 12. Frequency deviation of area 2 due to ΔP_{d2} = 0.05 p.u. with nonlinear effects

Figures 11-13 demonstrate the robustness of the proposed controller under a large dead zone width and GRC. In addition, it has been possible to reduce the steady state error in the tie line power flow deviations (Fig 10, 13) using the proposed controller while it has been difficult with a single layered FLC proposed by (Indulkar & Raj, 1995).

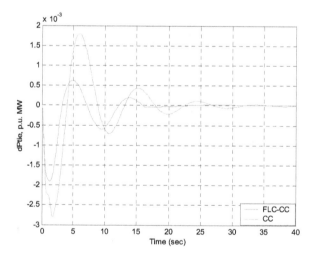

Figure 13. Tie line power deviation response due to ΔP_{d2} = 0.05 p.u.

The performance Evaluation of The proposed controller (FLC-CC) is given by table 3.

		t_s [sec]		OS [p.u.] $\cdot 10^{-3}$		US [p.u.] $\cdot 10^{-3}$	
		w/NE	NE	w/NE	NE	w/NE	NE
ΔP_{tie}	FLC-CC	10	17	0	0	1.2	3
(Δf_1)	FLC-FLC	25	25	2.5	2.5	3.5	3.7
Δf_1	FLC-CC	5	5	0	0.1	45	52
	FLC-FLC	30	32	4.0	5	17	18
Δf_2	FLC-CC	12	10	0	0.1	45	45
	FLC-FLC	30	30	4	4	20	20

Table 3. Performance Evaluation Of The Proposed Controller (FLC-CC)

7. Conclusion

In this paper a two layered controller with a fuzzy pre-compensator is used to damp the power system frequencies and tie line power error oscillation and track their errors to zero. The price to pay for changing a conventional controller into a FLC in order to obtain a two layered controller, where both layers are FLC, would be in the computation. FLC is driven by a set of control rules rather than by two constant proportional and/or integral gains. In our approach, simple tuning of the conventional controller parameters enables the easy and cheap implementation of the proposed controller. Extensive simulations for a single area and an interconnected systems with no reheat, reheat and hydro turbines, taking into account a number of practical aspects such as the loads and parameters disturbances and the nonlinear effects, have verified the validity of our scheme over the conventional controller. Therefore, the proposed controller should be preferred. Further research is based on finding the optimum tuning method for the conventional controller parameters.

Author details

Mavungu Masiala, Mohsen Ghribi* and Azeddine Kaddouri
Greater Research Group, University of Moncton, New-Brunswick, Canada

Appendix A

Nominal Parameters Of A Typical Single-Area Nonreheat Power System (Model – 1):

T_p - Electric system time constant = 20,0 [s]

K_p - Electric system static gain = 120 $\frac{[\text{Hz}}{\text{p.u.MW}^{-1}]}$

T_t - Turbine time constant = 0.30 [s]

T_g - Governor time constant = 0.08 [s]

R - Speed regulation due to = 2.40 [Hz
  governor action   p.u.MW^{-1}]

ΔP_d - load demand change [p.u. MW]

ΔP_g - incremental generation change [p.u. MW]

ΔX_E - incremental governor valve position change

Ki - integral control gain

s - the Laplace operator

* Corresponding Author

Appendix B

Nominal Parameters Of A Two-Area Reheat Hydrothermal Power System (Model – 2) :

i	- subscript referring to area	=	1,2
ΔX_{gi}	- incremental governor valve position change		
K_i	- integral control gain		
T_{12}	- synchronising coefficient		
ΔP_{tie_i}	- incremental change in tie-line power	[p.u. MW]	
ΔP_{gi}	- incremental generation change	[p.u. MW]	
ΔP_{di}	- load demand change	[p.u. MW]	
P_{ri}	- rated area power	[p.u. MW]	
T_r	- reheat time constant	=	10.0 [s]
T_g	- governor time constant	=	0.08 [s]
T_t	- turbine time constant	=	0.30 [s]
R_i	- speed regulation due to governor action	=	2.40 [Hz p.u. MW⁻¹]
B_i	- frequency bias constant	=	0.425 [p.u. MW/Hz]
D_i	- load-frequency constant	=	0.0083 [p.u. MW/Hz]
P_{tie-m}	- maximum tie-line power handling capacity	=	200 [MW]
T_w	- water starting time constant	=	1.0 [s]
T_R	- hydro governor time	=	5.0 [s]
T_1	constants	=	48.7 [s]
T_2		=	0.513 [s]
a_{12}	- rated area power constant	=	$-P_{r1}/P_{r2}$
K_r	- high pressure turbine power fraction	=	0.50

8. References

Benjamin, N.N & Chan, W.C. (1982). *Variable structure control of electric power generation*, IEEE Trans. Power Apparatus and Systems, Vol. PAS-101, No. 2, pp. 376 – 380

Pan C. T., & Liaw, C. M. (1989). *An Adaptive controller for power system load-frequency control*, IEEE Trans. Power Systems, (February 1989), Vol. 4, No. 1, pp. 122 – 128

Kothari, M. L.; Nanda, J.; Kothari, D. P. & Das, D. (1989). *Discrete-mode automatic generation control of a two-area reheat thermal system with new area control error*, IEEE Trans. Power Systems, (May 1989), vol. 4, No. 2, pp. 730 – 738

Wang, Y.; Zhou, Y. R. & Wen C. (1994). *New robust adaptive load-frequency control with system parametric uncertainties*, IEEE Trans. Generation, Transmission and Distribution, (May 1994), Vol. 141, No. 3, pp. 184 – 190

Indulkar, C. S. & Raj B. (1995). *Application of fuzzy controller to automatic generation control*, Electrical Machines and Power Systems, Vol. 23, pp. 209 – 220

Karnavas, Y.L. & Papadopoulos, D.P. (2002). *AGC for autonomous power system using combined intelligent techniques*, Electric Power Systems Research, 62, pp. 225 – 239

Moon, Y-H.; Ryu, H-S.; Lee, J-G.; Song, K-B. & Shin, M-C. (2002). *Extended integral control for load frequency control with the consideration of generation-rate constraints*, Electrical Power and Energy Systems, vol. 24, pp. 263 – 269

Sherbiny, M.K.; El-Saady, G. & Yousef, A.M. (2003). *Efficient fuzzy logic load-frequency controller*, Energy Conversion & Management, Vol. 43, pp. 1853 – 1863

Corradini, M. L. & Orlando, G. (2002). *Robust stabilization of nonlinear uncertain plants with backlash or dead–zone in the actuator*, IEEE Trans. Control Systems Technology, vol. 10, No. 1, pp. 158 – 166

Tao, G. and Kokotovic, P.V. (1994). *Adaptive control of plants with unknown dead–zones*, IEEE Trans. Automatic Control, (January 1994), Vol. 39, No. 1, pp. 59 – 68

Wang, X-S.; Su, C-Y. & Hong H. (2004). *Robust adaptive control of a class of nonlinear systems with unknown dead-zone*, Automatica, vol. 40, pp. 407 – 413

Lee, C. C. (1990). *Fuzzy logic in control systems : Fuzzy logic controller – Part I, II,"* IEEE Trans. Systems, Man and Cybernetics, (March-April 1990), Vol. 20, No. 2, pp. 404 – 435

Kim, J-H.; Park, J. H.; Lee, S-W. & Chong, E.K.P. (1994). *A two layered fuzzy logic controller for systems with dead zones*, IEEE Trans. Industrial Electronics, Vol. 41, No. 2, pp. 155–162

Yeh, Z-M. & Li, K-H. *A systematic approach for designing multistage fuzzy control systems*, Fuzzy Sets and Systems, Article in press, pp. 1 – 23, 2003.

Oh, S-Y. & Park, D-J. *Design of new adaptive fuzzy logic controller for nonlinear plants with unknown or time-varying dead zones*, IEEE Trans. Fuzzy Systems, vol. 6, No. 4, pp. 482 – 491, November 1998

Koo, K-M. *Stable adaptive fuzzy controller with time–varying dead–zone*, Fuzzy Sets and Systems, vol. 121, pp. 161 – 168, 2001

Rubai, A. *Transient stability control: A multi-level hierarchical approach*, IEEE Trans. Power Systems, vol. 6, No. 1, pp. 262 – 268, February 1991

Wang, Li-Xi (1997). *A course in fuzzy Systems and control*, Prentice Hall PTR, Upper Saddle River, NJ, pp. 249 – 257

Mamoh, J. A. and K. Tomsovic, K. (1995). *Overview and literature survey of fuzzy set theory in power systems*, IEEE Trans. Power Systems, (August 1995), Vol. 10, no. 3, pp. 1676-1690

Das, D. ; Kothari, M.L.; Kothari, D.P. and Nanda, J., (1991). *Variable structure control strategy to automatic generation control of interconnected reheat thermal system*, IEEE Proceedings on Control Theory and Applications, (November 1991), Vol. 138, no. 6, pp. 579-585

Djukanovic, M. (1995). *Conceptual development of optimal load frequency control using artificial neural networks and fuzzy set theory*, International Journal of Engineering Intelligent Systems for Electrical Engineering & Communication, vol. 3, no. 2. pp. 95-108, 1995.

Ha, Q. P. and Negnevitsky, M., (1997). *Fuzzy tuning in electric power system generation control*, Proceedings of the 4th Conference on APSCOM, (November 1997), vol. 2, no. 450, pp.662-667

Chang, C.S. and Fu, W. (1997). *Area load frequency control using fuzzy gain scheduling of PI controllers*, Electric Power Systems Research, Vol. 42, pp. 145-152

Yesil, E.; Guzelkaya, M. and Ekisin, I. (2004). *Self-tuning fuzzy PID type load and frequency controller*, Energy Conversion and Management, Vol. 45, pp. 377-390

Talaq J. and Al-Basri, F. (1999). *Adaptive fuzzy gain scheduling for load frequency control*, IEEE Trans. Power Systems, (February 1999), Vol. 14, no. 1, pp. 145-150

Li, H.-X., and Gatland, H.B. (1996). *Conventional fuzzy control and its enhancement*, IEEE Trans. Systems, Man and Cybernetics – Part B: Cybernetics, (October 1996), Vol. 26, No. 5, pp. 791-797

Chown, G.A. and Hartman, R.C. (1998). *Design and experiment with a fuzzy controller for automatic generation control (AGC), IEEE Trans. Power Systems,* (Aug. 1998), Vol. 13, pp. 965-970

Fosha, C.E. and Elgerd, O.I. (1970). *The megawatt frequency control problem: a new approach via optimal control theory,* IEEE *Trans. Power Apparatus and Systems,* (April 1970), Vol. PAS-89 (4), pp. 563-577

Passino, K.M. and Yourkovich, S. (1998), *Fuzzy Control,* Menlo Park, California: Addison Wesley Longman

Chapter 9

Discrete-Time Cycle-to-Cycle Fuzzy Logic Control of FES-Induced Swinging Motion

B. S. K. K. Ibrahim, M. O. Tokhi, M. S. Huq and S. C. Gharooni

Additional information is available at the end of the chapter

1. Introduction

Functional electrical stimulation (FES) can be used to restore motor function to individuals with spinal cord injuries (SCI). FES involves artificially inducing a current in specific motor neurons to generate a skeletal muscle contraction. FES induced movement control is a significantly challenging area for researchers. The challenge mainly arises due to muscle response characteristics such as fatigue, time-varying properties and nonlinear dynamics of paralyzed muscles [1]. Another challenge is due to certain motor reflexes such as spasticity. Spasticity is a reflex or uncontrolled response to something that excites the nerve endings and produces muscle contractions. These reflexes are often unpredictable and may impede joint movements [2].

Primarily due to the complexity of the system (nonlinearities, time-variation) practical FES systems are predominantly open-loop where the controller receives no information about the actual state of the system [3]. In its basic form, these systems require continuous user input. Practical success of this open-loop control strategy is still, however, seriously limited due to the fixed nature of the associated parameters. The problem arises especially due to the existing parameter variations (e.g., muscle fatigue), inherent time-variance, and strong nonlinearities present in the neuromuscular-skeletal system or the plant to be controlled. Besides, in such open-loop control approach, the actual movement is not assessed in real time and any mechanism of adapting the stimulation pattern in response to unforeseen circumstances such as external perturbations or muscle spasms is absent [4]. These prominent problems can be resolved by having a suitable closed-loop adaptive control mechanism. Such approach has several advantages over open-loop schemes, including better tracking performance and smaller sensitivity to the modeling errors, parameter variations, and external disturbances [5].

In controlling cyclical movement, one can try to follow pre-set joint angle trajectories. Although the trajectory-based closed-loop control has been developed but it has not been used yet in clinical FES gait because of difficulties in achieving accurate tracking performance [6]. Moreover, in the swing phase of gait, following exact trajectories is unimportant and inefficient, leading to fatigue due to large forces that must be exerted to precisely control the high inertia body segments [3]. For these reasons, cycle-to-cycle control method is expected to be an alternative to trajectory based closed-loop FES control. The cycle-to-cycle control delivers electrical stimulation in the form of open-loop control in each cycle without reference trajectory but it is still closed-loop control. In this control strategy, movement parameters at the end of each cycle are compared as in the desired set point, and the stimulation for the next cycle is adjusted on the basis of the error in the preceding cycle.

In fact, FES induced movements have traditionally been achieved through application of stimulus bursts rather than continuous tracking control. The burst of stimulus signal would drive the joint to its desired orientation through ballistic movement and thus traversing a trajectory defined purely by the physics of the segment combination [5]. The cycle-to-cycle control approach retains this basic mechanism of movement generation through stimulus burst and comes into action when the movement is repetitive or cyclical, through automatic adjustment of the burst parameters to maintain the desired target orientation at each cycle [7]. While the trajectory based closed-loop control for knee joint angle of paraplegic has been criticized for having poor tracking and oscillatory responses and even its inability to reach full knee extension angle [8]. the ability of cycle-to-cycle control approach to realize the target joint orientation has been demonstrated in experimental tests of controlling maximum knee extension angle [9] or hip joint range [7].

Researches as in [7] and [9] investigated a discrete-time proportional-integral-derivative (PID) feedback controller for cycle-to-cycle adaptation of an experimentally initialized stimulation signal with a view to compensate for the fatigue-induced time variation of muscle output. For practical use of cycle-to-cycle control, realization in multi-joint control is crucial, in which problems seen in the PID controller such as drawback in determination of controller parameter values and a lack of capability in compensating muscle fatigue [7] have to be solved. It is difficult to establish the control parameters for these systems since they are not always the same under different circumstances [10]. Therefore, traditional control approaches, such as PID control might not perform satisfactorily if the system to be controlled is of highly nonlinear and uncertain nature [5].

On the other hand, fuzzy logic control (FLC) has long been known for its ability to handle a complex nonlinear system without developing a mathematical model of the system. FLC is the fastest growing soft computing tool in medicine and biomedical engineering [11]. It is being used successfully in an increasing number of application areas in the control community. FLCs are rule-based systems that use fuzzy linguistic variables to model human rule-of-thumb approaches to problem solving, and thus overcoming the limitations that classical expert systems may face because of their inflexible representation of human decision making. The major strength of fuzzy control also lies in the way a nonlinear output mapping of a number of inputs can be specified easily using fuzzy linguistic variables and fuzzy rules [12]. The control signal is computed by rule evaluation called fuzzy inference instead of by mathematical equations. In order to compensate the non-linearity of the musculo-skeletal system responses, the cycle-to-cycle control was implemented using fuzzy controller. Thus FLC with is the preferred option in the current work

This chapter presents the development of strategies for swinging motion control by controlling the amount of stimulation pulsewidth to the quadriceps muscle of the knee joints. The capability of the controller to control knee joint movements is first assessed in computer simulations using a musculo-skeletal knee joint model. The knee joint model developed in Matlab/Simulink, as described in [13], is used to develop an FLC-based cycle-to-cycle control strategy for the knee joint movement. The FLC output is the controlled FES stimulation pulsewidth signal which stimulates the knee extensors providing torque to the knee joint. The swinging movement is performed by only controlling stimulation pulsewidth to the knee extensors to extent the knee and then the knee is left freely to flex in the flexion period. The controllers are then tested through experimental work on a paraplegic in terms of swinging performance and compensation of muscle fatigue and spasticity.

2. Materials and method

2.1. Model of knee joint

The shank-quadriceps dynamics are modelled as the interconnection of passive and active properties of muscle model and the segmental dynamics. The total knee-joint moment is given as [14]:

$$M_i = M_a + M_g + M_s + M_d \tag{1}$$

where M_i refers to inertial moment, M_g is gravitational moment, M_a refers to an active knee joint moment produced by electrical stimulation, M_s is the knee joint elastic moment and M_d is the viscous moment representing the passive behaviour of the knee joint. In this research the M_i and M_g are represented by the equations of motion for dynamic model of the lower limb while M_a and M_s+M_d are represented by a fuzzy model as active properties of quadriceps muscle and passive viscoelasticity respectively. A schematic representation of the knee joint model consisting of active properties, passive viscoelasticity and equations of motion of the lower limb is shown in Figure 1. The active joint moment is added with the passive joint moment as an input (torque) to the lower limb model and this will produce the knee angle as the output. The subject participating in this work was a 48 year-old T2&T3 incomplete paraplegic male with 20 years post-injury with height = 173cm and weight = 80kg. Informed consent was obtained from the subject.

A schematic diagram of the lower limb model is shown in Figure 5, where q_2 = shank length, r_1 = position of COM along the shank, r_2 = position of COM along the foot, θ_1 =knee angle and θ_2 =ankle angle. Hence, the dynamics of motion can be represented in the simpler form based on Kane's equations as in [13]. The gravitational (M_g) moment is represented by:-

$$M_g = m_1 g \cos\theta_1 r_1 + m_2 g \cos\theta_1 q_2 \qquad (2)$$

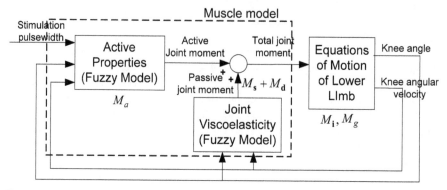

Figure 1. Schematic representation of the knee joint model

The inertial (M_i) moment of the lower limb is represented as follows:-

$$M_i = -m_2 q_2 \dot{\theta}_1^2 r_2 - I_1 \ddot{\theta}_1 - m_1 r_1^2 \ddot{\theta}_1 - m_2 q_2^2 \ddot{\theta}_1 \qquad (3)$$

where, m_1 = shank mass, m_2 =foot mass, I_1= moment of inertia about COM, $\dot{\theta}_1$ =knee velocity, $\ddot{\theta}_1$ =knee acceleration, g =gravity=9.81 m/s².

Anthropometric measurements of length of the lower limb were made and this is shown in Table 1.

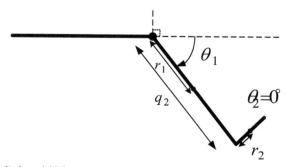

Figure 2. Lower limb model [4]

The knee joint model input is the stimulation pulsewidth as would be delivered in practice by an electrical stimulator. The complete model of knee joint thus developed is utilized as platform for simulation of the system and development of control approaches.

Segment	Length (m)
Shank length	0.426
Foot length	0.068
Approximated position of COM of shank	0.213
Approximated position of COM of soot	0.034

Table 1. Anthropometric data of subject

2.2. Cycle to cycle controller development

Researcher as in [15] highlighted cycle-to-cycle control as a method for using feedback to improve product quality for processes that are inaccessible within a single processing cycle but can be changed between cycles. The same concept has been applied in this study, where only reaching a target joint orientation through ballistic movement is taken into consideration rather than rigorously following a trajectory. The muscle is stimulated by single burst of controlled stimulation pulsewidth for each cycle to induce joint movement reaching the target extension knee angle. Therefore, the method is different from the traditional closed-loop control such as tracking control of desired angle trajectory.

An outline of the discrete-time fuzzy control based cycle-to-cycle control is shown in Figure3. The controlled maximum joint angle of the previous cycle is delivered as feedback signal. Error is defined as difference between the target and measured joint angle. The controller will regulate the duration of stimulation pulsewidth based on the error and previous flexion angle.

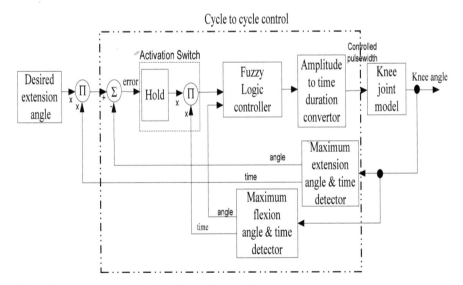

Figure 3. Discrete-time FLC based cycle to cycle control

2.2.1. Maximum flexion and extension detector

The maximum flexion time detector will detect the time the angle reaches the peak of knee flexion for each cycle. The maximum extension signal and time detector will detect the peak angle of knee extension and the time the knee angle reaches this point for each cycle. The extension and flexion stages of the knee angle are shown in Figure 4.

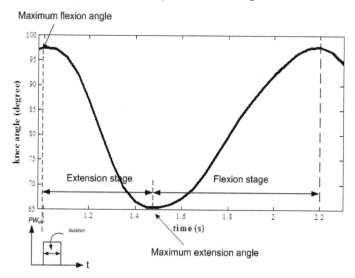

Figure 4. Extension and flexion stages

2.2.2. Activation switch

Activation switch consists of hold and multiplier block to be active only when the knee angle reaches maximum flexion. The activation switch will hold the error signal and produce the output whenever it receives a signal from the maximum flexion time detector.

2.2.3. Amplitude to time duration convertor

Amplitude to time duration converter is linearly converting the controlled signal (amplitude) from controller to time duration using signal comparator and shifting technique as shown in Figure 6.3. In this technique, first the controlled signal is compared with specific constant values for low to high in the parallel structure. Each comparator compares the controlled signal with the specific constant, if the controlled signal is greater than or equal to the specific constant then a single pulse will pass through the comparator. The first comparator compares the control signal with zero, if there is any signal from controller then the output will be a pulse with 0.05s width. The second comparator compares the control signal with specific constant and shift 0.05s and the next comparator compares and shifts by a further 0.05s. Then the resultant pulse duration for each cycle is obtained by summing up the total pulses passed through the comparators and amplifying the signal with 220µs. The higher the controlled signal the more gates can be passed through and the wider the duration of pulse. Therefore the output of this converter is a single burst of controlled stimulation pulse duration with constant amplitude (220µs) for each cycle.

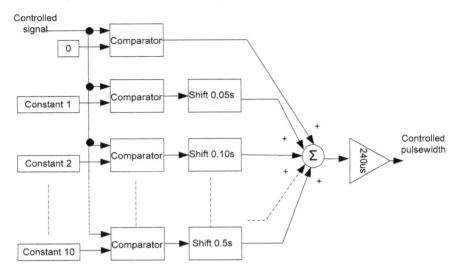

Figure 5. Signal comparator and shifting technique

2.2.4. Controller objectives

The FLC-based cycle-to-cycle control was designed to achieve the following objectives:

i.   Able to reach full knee extension
ii. Able to reach target extension angle thus maintain a steady swinging motion.
iii. Compensate for muscle fatigue
iv. Compensate for spasticity

2.2.5. Fuzzy controller design

Measured output of the controlled musculoskeletal system of the previous cycle is delivered as feedback signal. Proper value of signal is determined and regulated automatically by a Sugeno-type fuzzy controller using control rules as shown in Table 2. Input membership function is expressed as triangle fuzzy sets. Output membership function is expressed as fuzzy singletons. Input of fuzzy controller is aggregated by fuzzy inference using fuzzy rules to produce control action.

The fuzzy rules base directs control action based on error and flexion angle. The error will be higher when the muscle fatigues, in which case the response to a stimulation burst will change. To compensate for this changing system response, the stimulation burst time has to be increased such that shank can reach the desired angle in every cycle. The flexion angle was taken into account to rule out the disturbance due to spasticity. Combination of the information about error and knowledge about flexion angle will be necessary for controller to give an appropriate stimulation pulsewidth in compensation of muscle fatigue and motor reflexes. pulsewidth in compensation of muscle fatigue and motor reflexes.

		Flexion angle					
		Very High	High	Normal	Low	Very Low	Extremely Low
Error	Negative	Low	Low	Low	Low	Zero	Zero
	Zero	Low	Low	Low	Low	Zero	Zero
	Positive Small	Low	Normal	Normal	Very Low	Very Low	Zero
	Positive Medium	Low	Normal	Normal	Normal	Very Low	Zero
	Positive Big	Normal	Normal	Normal	Normal	Normal	Zero

Table 2. Rules of Sugeno-type FLC

3. Results and discussion

This discrete-time fuzzy logic cycle to cycle control technique emphasizes the view to overcome some drawbacks of trajectory based closed-loop FES control such as poor tracking, oscillating response and inability to reach full knee extension angle (Hatwell, 1991). Then the capabilities in compensating for muscle fatigue and spastisity are investigated. The ability of this control approach to realize the target joint orientation is assessed in simulation and experimental test as follows:-

3.1. Controllers' performance in simulation environment

A complete set of non-linear dynamic equations of the knee joint model comprising the passive properties and active properties have been used in the simulations for purposes of controller development. Computer simulations are performed to assess the performance of the designed discrete-time fuzzy logic cycle-to-cycle control approach in generating stimulation burst durations for the desired extension angle. The simulations were carried out within the Matlab/Simulink environment. The muscle model was controlled by changing the pulse width; however the amplitude and the frequency of the stimulation pulses were constant. Here only the knee extensors are controlled by applying regulated stimulation pulsewidth to the quadriceps muscle model.

3.1.1. Full knee extension angle

The test was initiated with stimulation pulse of 240µs amplitude with 0.3s burst duration for the first cycle of swing in gait before activating the controller. FES induced swinging motion was controlled using fuzzy controller to reach full extension angle. The full extension angle that can be achieved by paraplegic was defined as 10°. The computer simulation test was performed with stimulation course of 50 cycles. The first 5 cycles of the controlled swinging leg test of the full extension knee angle are shown in Figure 6.4. As can be seen, using FLC-based cycle-to-cycle control approach the first objective was achieved; to reach the full knee extension. The knee reached full extension at 3rd cycle and was able to maintain the swinging motion without any predefined trajectory.

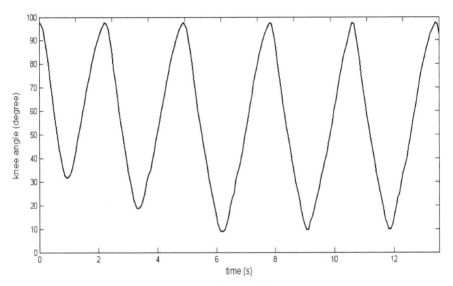

Figure 6. Controlled swinging leg for desired angle at 10° (full extension)

3.1.2. Target extension angle and maintain steady swinging

The desired extension knee angle was set to be at 65° as considered in [16]. The test was initiated with stimulation pulse of 220µs amplitude with 0.25s burst duration for the first cycle before activating the controller.

3.1.2.1. Without presence of musle fatique and voluntary activation

The test is to achieve the target extension angle and thus maintain a steady swing of the shank without presence of musle fatique and voluntary activation. In the each test computer simulation was performed with stimulation course of 50 cycles. The first 10 cycles of the controlled swinging leg test of the knee joint at 65° is shown in Figure 7. As can be seen the cycle-to-cycle control approach can achieve the target extension angle at 3rd and thus maintain a steady swing of the shank. It is noted that the performance of the controller was quite good and acceptable.

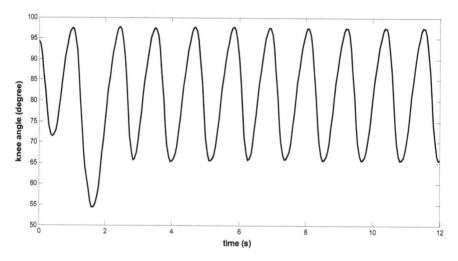

Figure 7. Controlled swinging leg for desired angle at 65° (normal extension)

3.1.2.2. With presence of musle fatique and without presence of voluntary activation

Muscle fatigue is an inevitable pitfall in FES induced control of movements. Fatigue is defined as the inability of a muscle to continue to generate a required force. It limits the duration FES can be effective by drastically reducing the muscle force output. It thus should be considered as an important criterion for FES induced movement control. The fatigue resistance of the control approaches were analyzed based on relative drop in knee extension by simulating the fatigue as in Figure 8. The fatigue simulation was to reduce muscle torque output to the 80% of total torque output at the end of simulation [5]. Figure 6.7 shows the knee joint response obtained using this controller, but with FES torque dropped down to 80% of its normal value. The effect of fatigue can be noted at 3rd cycle, the shank was unable to reach the target angle. Then controller has taken action to overcome this by increasing the stimulation burst time. Then, after 6th cycle the shank reached the target angle thus maintaining the swinging motion. As can be seen the controller performed very well in terms of robustness of the FLC in the presence of muscle fatigue.

3.1.2.3. With presences of musle fatique and spasticity

SCI muscle often exhibits spasticity [17] and may vary with time during cyclical motion [18]. Spasticity is defined as motor disorder characterized by a velocity-dependent increase in tonic stretch reflexes with exaggerated tendon jerks [19]. With regard to FES-approaches for function restoration, spasticity certainly represents a disadvantage especially flexion spasticity [20]. Muscle fatigue (as in previous test) and motor reflex to represent spasticity were simulated to assess the ability of controller to tackle these influences.

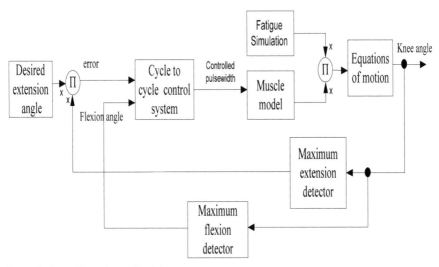

Figure 8. Controller with simulated fatigue

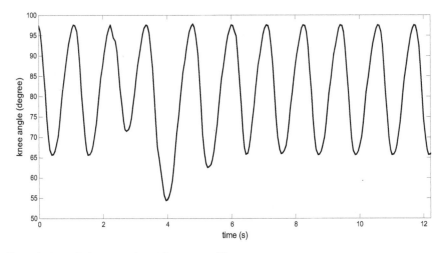

Figure 9. Controlled swinging leg with presence of fatigue

Since in practice spasticity is unpredictable, the motor reflex was simulated by multiplying a random number (between 0.1 to 1) with flexion angle as the second controller's input as in Figure 10. Controller has to compensate for the presence of spasticity in order to maintain swinging motion. Figure 11 shows the knee joint response obtained using this controller with the influence of muscle fatigue and voluntary activation. The presence of spasticity can be noted at 3rd cycle and for almost 5sec, then shank returning back to rest angle. When severe spasticity happens, the controller will stop the stimulation because it may seriously hinder the swinging activity as well as for reason of safety. Once knee angle reaches the rest angle, the controller starts the stimulation. However, as can be seen in Figure 11 muscle spasm is noted leading to muscle fatigue. The controller was able to compensate for the presence of fatigue after few cycles. The controller performed well in terms of robustness of the FLC overcoming the influence of these phenomena after few cycles.

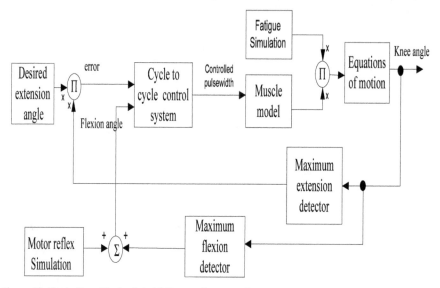

Figure 10. Controller with simulated fatigue and motor reflex

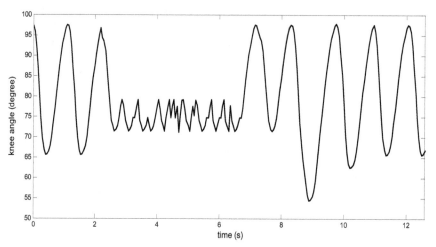

Figure 11. Controlled swinging leg with influence of fatigue and voluntary activation

3.2. Experimental validation of controller

The laboratory apparatus built to study the knee joint control by FES is shown in Figure 12. The subject sat on a chair, which allowed the lower leg to swing freely, while the ankle angle was fixed at 0°. The knee extensors (quadriceps muscle group) were stimulated by a pair of surface electrodes (2"x5"). The cathode was placed on the motor point of rectus femoris and the anode was placed distally at the quadriceps tendon. Knee angle was defined in Figure 2, with when the lower leg was at rest during knee flexion (i.e., 90).

The computer-controlled stimulator system consisted of a personal computer, computer-controlled interface (including analog-to-digital converter), current controlled stimulator and electro-goniometer (see Figure 12). The stimulation pulsewidth is generated by FLC based on the error by comparing the actual extension angle and the desired ones. All these operations were performed in the Matlab/Simulink environment in the computer. The Hasomed stimulator device was connected to PC via USB interface port. The knee joint angle was measured via the Biometric flexible electroganiometer mounted at approximate center of rotation of the knee joint. Stimulation pulsewidth ranged from 0 to 230μs and stimulation current was fixed to 40mA with a biphasic type pulse. The stimulation frequency was set to 25Hz and the knee joint angle sampling time was 0.05s. The experimental validation tests of the discrete-time fuzzy logic cycle to cycle control (based on simulation study) was assessed the capability of the controller to control the swinging motion as desired.

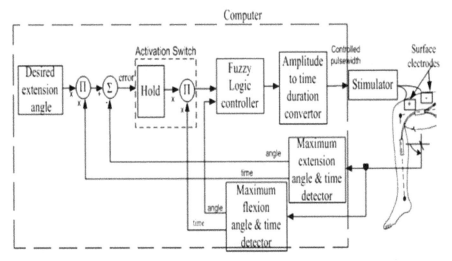

Figure 12. The equipment setup of this study

3.2.1. Full knee extension angle

The test was initiated with stimulation pulse of 240μs amplitude with 0.3s burst duration for the first cycle of swing in gait before activating the controller as in the simulation study.

FES-induced swinging motion was controlled using fuzzy controller to reach full extension angle at 10°. The shank was able to reach full knee extension at 4th cycle as can be seen in Figure 13. The ability of the cycle-to-cycle control approach to reach full knee extension has been demonstrated similar to that in [9].

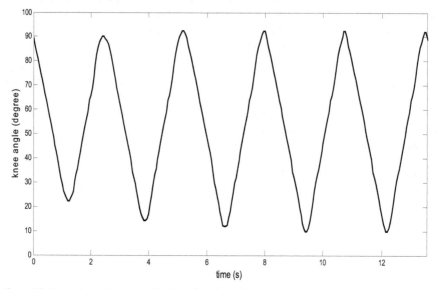

Figure 13. Controlled swinging leg for desired angle at 10° (full extension)

3.2.2. Target extension angle and maintaining steady swinging

The same procedure as in the simulation work was applied with the test initiated with 220µs amplitude with 0.15s burst durations of stimulation pulse. The controller performed in high intense stimulation course of 100 cycles as shown in Figure 14 in order to get influences of muscle fatigue and spasticity. EMG signal via surface electrodes on quadriceps muscle was recorded in this test to monitor EMG activity. The controller was tested in three scenarios, as in the simulations. Few trials were conducted in order to make sure the presence of these three scenarios was on the same stimulation course. An intra-trial interval for 120s was used to reduce the effect of fatigue in the beginning of the stimulation. Finally the best stimulation course with presence of both phenomena; muscle fatigue and spasticity were recorded. The recorded stimulation course was for almost 100 cycles and this was divided into three scenarios as follows:

3.2.2.1. Without presence of musle fatique and voluntary activation

In the first part of test, the controller was validated without fatigue and spasticity by considering only the first 10 cycles of the stimulation course. Before beginning the test, the subject was asked to relax as much as possible. There was no EMG present when the subject was fully relaxed. Figure 14 shows the response of knee angle in the first scenario. The controller was able to perform a steady swinging motion of shank after 5s with ability to extend the knee to the desired extension angle. Hence this controller achieved the main objective; to maintain a steady swinging of the lower limb as desired but the controller needed more time to achieve this.

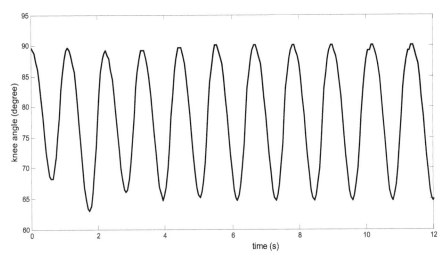

Figure 14. Controlled swinging leg (Experimental work)

3.2.2.2. With presence of musle fatique and without presence of voluntary activation

In the second part of test, the controller was validated when the muscle fatigue happened due to high stimulation intensity. This scenario can be seen by monitoring the reduction in the extension of knee joint. Few cycles of the stimulation course before and after fatigue were considered in this test as shown in Figure 15. It can been seen that at the beginning the knee angle was unable to reach the desired value due to fatigue, and then with the controller action by increasing the stimulation burst time the shank was able move a bit higher. After few cycles the knee angle reached the desired extension angle and swinging motion was maintained. Therefore, this controller achieved the third objective; to maintain a steady swinging of the lower limb as desired in presence of muscle fatigue. Only a small amount of EMG activity was recorded in this scenario.

3.2.2.3. With presence of musle fatique and voluntary activation

In the final validation test, the controller was validated with presence of spasticity and muscle fatigue. This scenario can be seen by monitoring the knee angle at which spasticity stops the natural swing. Furthermore, high EMG activity was apparent throughout the trace. These uncontrolled muscle movements were brought by spasms causing fatigue. Thus, the reduction in the extension of knee joint can be seen after spasm. 15 cycles of the stimulation course before and after spasm were considered in this test as shown in Figure

16. It can be seen that spasticity caused uncontrolled movement. The controller automatically stopped the stimulation due to spasticity. Once there was no motion of the knee, the controller started the stimulation with widest stimulation burst time to compensate for muscle fatigue. After 5s the knee angle reached the desired extension angle and swinging motion was maintained. Therefore, this controller achieved the last objective; to maintain a steady swinging of the lower limb as desired in presence of spasticity and muscle fatigue.

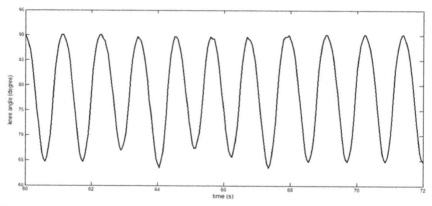

Figure 15. Controlled swinging leg with presence of fatigue (Experimental work)

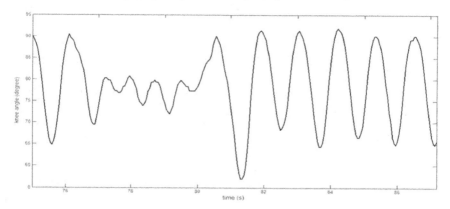

Figure 16. Controlled swinging leg with influence of fatigue and voluntary activation (Experimental work)

4. Conclusion

FES induced movement control is a difficult task due to the highly time-variant and nonlinear nature of the muscle and segmental dynamics. The great merit of a musculoskeletal model of knee joint is to serve for control development. In this study, a new closed-loop control approach using fuzzy logic based cycle-to-cycle control for FES-induced motion control has been proposed. This control technique also emphasizes to overcome some drawbacks of the trajectory based closed-loop FES control. The objectives of this controller have been set to achieve full knee extension angle, to reach target extension angle thus maintain a steady swinging motion and to compensate for muscle fatigue and spasticity. The performance of the controller to achieve these objectives has been assessed through simulation study and validated through experimental work. The controller has been proved to achieve all these objectives. Besides its suitability in generating the target joint orientation, one of the attractions of the cycle-to-cycle control in FES application is the absence of any reference trajectory. Cycle-to-cycle control is easy to implement in practice. Additionally, this method can compensate for non-linearity and time-variance of response of electrically stimulated musculoskeletal system. This controller may be suitable not only for swinging but also for other FES control applications involving movement of cyclical nature.

Author details

B. S. K. K. Ibrahim
Dept of Mechatronics and Robotics, Faculty of Electrical & Electronic Engineering, University Tun Hussein Onn Malaysia, Batu Pahat, Johor, Malaysia

M. O. Tokhi, M. S. Huq and S. C. Gharooni
Department of Automatic Control and System Engineering, University of Sheffield, United Kingdom

Acknowledgement

B.S.K.K.Ibrahim greatly acknowledges the financial supports of Higher Education Ministry of Malaysia and University Tun Hussein Onn, Malaysia.

5. References

[1] M. Levy, J. Mizrahi, and Z. Susak, "Recruitment force and fatigue characteristicsof quadriceps muscles of paraplegics isometrically activated by surface functional electrical stimulation," J. Biomed. Eng., vol. 12, pp. 150–156, 1990.
[2] CL Lynch and MR Popovic Closed-Loop Control for FES: Past Work and Future Directions, 10th Annual Conference of the International FES Society, July 2005 – Montreal, Canada.
[3] P.E. Crago,N. Lan, P. H. Veltink, J.J. Abbas and C.Kantor. New control strategies for neuroprosthetic systems, Journal Rehabilitation Res Device 33(2):158-72,1996.
[4] J. J. Abbas, "Feedback control of coronal plane hip angle in paraplegic subjects using functional neuromuscular stimulation," IEEE Trans. Biomed. Eng., vol. 38, pp. 687–698, 1991.

[5] M.S.Huq, Analysis and control of hybrid orthosis in therapeutic treadmill locomotion for paraplegia, PhD Thesis. The University of Sheffield, Sheffield, UK, 2009.

[6] A. Arifin, T. Watanabe, and N. Hashimiya, (2002). A Test of fuzzy controller for cycle-cycle control of FES-induced hemiplegics gait: Computer simulatilm in single-joint control: Proc. 36th Control of Japanese Soc.& Med. & Bid. Eng., Tohoku Chapter, p.30.

[7] H. M. Franken, P. H. Veltink, R.Tijsmans , H. Nijmeijer, , and H. B. K. Boom, (1993). Identification of passive knee joint and shank dynamics in paraplegics using quadriceps stimulation, IEEE Transactions on Rehabalitation Engineering, vol. 1,pp. 154-164.

[8] M.S. Hatwell, B.J. Oderkerk, C.A. Sacher, and G.F. Inbar,(1991). The development of a model reference adaptive controller to control the knee joint of paraplegics. IEEE Transactions on Automatic Control 36 6, pp. 683–691.

[9] P. H. Veltink. Control of FES-induced cyclical movements of the lower leg. Med Biol Eng Comput 1991 :29:NS8-12.

[10] Y.L. Chen, S.C. Chen, W.L. Chen, C.C. Hsiao, T.S. Kuo and J.S. Lai, Neural network and fuzzy control in FES-assisted locomotion for the hemiplegic, J Med Eng Technol 28 (2004), pp. 32–38.

[11] A. Yardimci, A Survey on Use of Soft Computing Methods in Medicine, Artificial Neural Networks – ICANN 2007,Springer Berlin / Heidelberg, pp. 69-79, 2007.

[12] T.C. Chin, and X. M. Qi, Genetic Algorithms for learning the rule base of fuzzy logic controller. Fuzzy Sets and Systems 97:1, 1998.

[13] B.S. K. K. Ibrahim, M.O. Tokhi M.S. Huq, and S.C. Gharooni, An Approach for Dynamic Characterisation of Passive Viscoelasticity and Estimation of Anthropometric Inertia Parameters of Paraplegic Knee Joint, unpublished.

[14] M. Ferrarin, and A. Pedotti. The relationship between electrical stimulus and joint torque: a dynamic model. IEEE Transactions on Rehabilitation Engineering, vol. 8 (3), pp. 342-352, 2000.

[15] A.K. Rzepniewski, Cycle-to-cycle control of multiple input-multiple output manufacturing processes, Ph.D. Thesis in Mechanical Engineering, Massachusetts Institute of Technology, Cambridge (2005).

[16] Fahey, T.D., Harvey, M., Schroeder, R. and Ferguson, F. (1985) Influence of sex differences and knee joint position on electrical stimulation-modulated strength increases. Medicine Science Sports Exercise 17, 144-147.

[17] Bobet, J. and Stein, R.B. (1998). A simple model of force generation by skeletal muscle during dynamic isometric contractions. IEEE Trans Biomed Eng, 45(8):1010-1016.

[18] Yang, J., F. Fung, J., Edamura, M. et al., (1991). H-reflex modulation during walking in spastic paretic subjects. Canadian Journal of Neurological Sciences 18(4):443–452.

[19] Lance JW. Symposium synopsis. In: Feldmann RG, Young RR, Koella WP, eds.Spasticity: disordered motor control. Chicago: Year Book Medical Publishers, 1980:485–95.

[20] Kralj, A. and Bajd, T, Functional Electrical Stimulation, Standing and Walking After Spinal Cord Injury CRC Press, Boca Raton, FL, 1989.

Chapter 10

Three Types of Fuzzy Controllers Applied in High-Performance Electric Drives for Three-Phase Induction Motors

José Luis Azcue, Alfeu J. Sguarezi Filho and Ernesto Ruppert

Additional information is available at the end of the chapter

1. Introduction

The electric drives are very common in industrial applications because they provide high dynamic performance. Nowadays exist a wide variety of schemes to control the speed, the electromagnetic torque and stator flux of three-phase induction motors. However, control remains a challenging problem for industrial applications of high dynamic performance, because the induction motors exhibit significant nonlinearities. Moreover, many of the parameters vary with the operating conditions. Although the Field Oriented Control (FOC) [16] schemes are attractive, but suffer from a major disadvantage, because they are sensitive to motor parameter variations such as the rotor time constant, and an incorrect flux estimation at low speeds. Another popular scheme for electric drives is the direct torque control (DTC) scheme [15][8], and an another DTC scheme based on space vector modulation (SVM) technique that reduces the torque ripples. This scheme does not need current regulators because its control variables are the electromagnetic torque and the stator flux. In this chapter we use the DTC-SVM scheme to analyze the performance of our proposed fuzzy controllers.

In the last decade, there was an increasing interest in combining artificial intelligent control tools with conventional control techniques. The principal motivations for such a hybrid implementation were that fuzzy logic issues such as uncertainty (or unknown variations in plant parameters and structure) can be dealt with more effectively. Hence improving the robustness of the control system. Conventional controls are very stable and allow various design objectives such as steady state and transient characteristics of a closed loop system. Several [5][6] works contributed to the design of such hybrid control schemes.

However, fuzzy controllers, unlike conventional PI controllers do not necessarily require the accurate mathematic model of the process to be controlled; instead, it uses the experience and knowledge about the controlled process to construct the fuzzy rules base. The fuzzy logic controllers are a good alternative for motor control systems since they are well known for treating with uncertainties and imprecisionï£¡'s. For example, in [1] the PI and fuzzy logic controllers are used to control the load angle, which simplifies the induction motor drive system. In [7], the fuzzy controllers are used to dynamically obtain the reference voltage vector in terms of torque error, stator flux error and stator flux angle. In this case, both torque and stator flux ripples are remarkably reduced. In [10], the fuzzy PI speed controller has a better response for a wide range of motor speed and in [3] a fuzzy self-tuning controller is implemented in order to substitute the unique PI controller, present in the DTC-SVM scheme. In this case, performance measures such as settling time, rise time and ITAE index are lower than the DTC-SVM scheme with PI controller.

The fuzzy inference system can be used to modulate the stator voltage vector applied to the induction motor [18]. In this case, unlike the cases mentioned above, the quantity of available vectors are arbitrarily increased, allowing better performance of the control scheme and lower levels of ripple than the classic DTC. However, it requires the stator current as an additional input, increasing the number of input variables. In this chapter we design and analyze in details three kinds of fuzzy controllers: the PI fuzzy controller (PI-F), the PI-type fuzzy controller (PIF) and the self-tuning PI-type fuzzy controller (STPIF). All of these fuzzy controllers are applied to a direct torque control scheme with space vector modulation technique for three-phase induction motor. In this DTC-SVM scheme, the fuzzy controllers generate corrective control actions based on the real torque trend only while minimizing the torque error.

2. Background

2.1. The three-phase induction motor dynamical equations

By the definitions of the fluxes, currents and voltages space vectors, the dynamical equations of the three-phase induction motor in stationary reference frame can be put into the following mathematical form [17]:

$$\vec{u}_s = R_s\vec{i}_s + \frac{d\vec{\psi}_s}{dt} \tag{1}$$

$$0 = R_r\vec{i}_r + \frac{d\vec{\psi}_r}{dt} - j\omega_r\vec{\psi}_r \tag{2}$$

$$\vec{\psi}_s = L_s\vec{i}_s + L_m\vec{i}_r \tag{3}$$

$$\vec{\psi}_r = L_r\vec{i}_r + L_m\vec{i}_s \tag{4}$$

Where $\vec{u}_s$ is the stator voltage space vector, $\vec{i}_s$ and $\vec{i}_r$ are the stator and rotor current space vectors, respectively, $\vec{\psi}_s$ and $\vec{\psi}_r$ are the stator and rotor flux space vectors, ω_r is the rotor angular speed, R_s and R_r are the stator and rotor resistances, L_s, L_r and L_m are the stator, rotor and mutual inductance respectively.

The electromagnetic torque t_e is expressed in terms of the cross-vectorial product of the stator and the rotor flux space vectors.

$$t_e = \frac{3}{2}P\frac{L_m}{L_rL_s\sigma}\vec{\psi}_r \times \vec{\psi}_s \tag{5}$$

$$t_e = \frac{3}{2}P\frac{L_m}{L_rL_s\sigma}|\vec{\psi}_r|\,|\vec{\psi}_s|\sin(\gamma) \tag{6}$$

Where γ is the load angle between stator and rotor flux space vector, P is a number of pole pairs and $\sigma = 1 - L_m^2/(L_s L_r)$ is the dispersion factor.

The three-phase induction motor model was implemented in MATLAB/Simulink as is shown in [2].

2.2. The principle of direct torque control

In the direct torque control if the sample time is short enough, such that the stator voltage space vector is imposed to the motor keeping the stator flux constant at the reference value. The rotor flux will become constant because it changes slower than the stator flux. The electromagnetic torque (6) can be quickly changed by changing the angle γ in the desired direction. This angle γ can be easily changed when choosing the appropriate stator voltage space vector.

For simplicity, let us assume that the stator phase ohmic drop could be neglected in (1). Therefore $d\vec{\psi}_s/dt = \vec{u}_s$. During a short time Δt, when the voltage space vector is applied it has:

$$\Delta \vec{\psi}_s \approx \vec{u}_s \cdot \Delta t \tag{7}$$

Thus the stator flux space vector moves by $\Delta \vec{\psi}_s$ in the direction of the stator voltage space vector at a speed which is proportional to the magnitude of the stator voltage space vector. By selecting step-by-step the appropriate stator voltage vector, it is possible to change the stator flux in the required direction.

3. Direct torque control scheme with space vector modulation technique

In Fig. 1, we show the block diagram for the DTC-SVM scheme [14] with a Fuzzy Controller, the fuzzy controller will be substitute for the three kind of proposed Fuzzy Controller one for time. The DTC-SVM scheme is an alternative to the classical DTC schemes [15], [8] and [9]. In this one, the load angle γ^* is not prefixed but it is determinate by the Fuzzy Controller. Equation (6) shows that the angle γ^* determines the electromagnetic torque which is necessary to supply the load. The three proposed Fuzzy Controllers determine the load angle using the torque error e and the torque error change Δe. Details about these controllers will be presented in the next section. Figure 1 shows the general block diagram of the DTC-SVM scheme, the inverter, the control signals for three-phase two-level inverter is generated by the DTC-SVM scheme.

3.1. Flux reference calculation

In stationary reference frame, the stator flux reference $\vec{\psi}_s^*$ can be decomposed in two perpendicular components ψ_{ds}^* and ψ_{qs}^*. Therefore, the output of the fuzzy controller γ^* is added to rotor flux angle $\angle \vec{\psi}_r$ in order to estimate the next angle of the stator flux reference.

In this chapter we consider the magnitude of stator flux reference as a constant. Therefore, we can use the relation presented in equation (8) to calculate the stator flux reference vector.

$$\vec{\psi}_s^* = |\vec{\psi}_s^*| \cos(\gamma^* + \angle \vec{\psi}_r) + j |\vec{\psi}_s^*| \sin(\gamma^* + \angle \vec{\psi}_r) \tag{8}$$

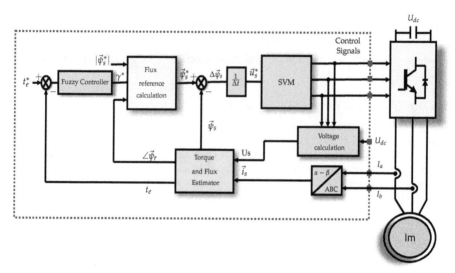

Figure 1. Direct torque control with space vector modulation scheme and fuzzy controller.

Moreover, if we consider the stator voltage $\vec{u}_s$ during a short time Δt, it is possible to reproduce a flux variation $\Delta\vec{\psi}_s$. Notice that the stator flux variation is nearly proportional to the stator voltage space vector as seen in the equation (7).

3.2. Stator voltage calculation

The stator voltage calculation uses the DC link voltage (U_{dc}) and the inverter switch state (S_{Wa}, S_{Wb}, S_{Wc}) of the three-phase two level inverter. The stator voltage vector $\vec{u}_s$ is determined as in [4]:

$$\vec{u}_s = \frac{2}{3}\left[\left(S_{Wa} - \frac{S_{Wb} + S_{Wc}}{2}\right) + j\frac{\sqrt{3}}{2}(S_{Wb} - S_{Wc})\right]U_{dc} \tag{9}$$

3.3. Electromagnetic torque and stator flux estimation

As drawn by Fig. 1 the electromagnetic torque and the stator flux estimation depend on the stator voltage and the stator current space vectors,

$$\vec{\psi}_s = \int(\vec{u}_s - R_s \cdot \vec{i}_s)dt \tag{10}$$

The problem with this kind of estimation is that for low speeds the back electromotive force (emf) depends strongly of the stator resistance, to resolve this problem is used the current model to improve the flux estimation as in [13]. The rotor flux $\vec{\psi}_{rdq}$ represented in the rotor flux reference frame is given by:

$$\vec{\psi}_{rdq} = \frac{L_m}{1 + sT_r}\vec{i}_{sdq} - j\frac{(\omega_{\psi_r} - \omega_r)T_r}{1 + sT_r}\vec{\psi}_{rdq} \tag{11}$$

Notice that $T_r = L_r / R_r$ is the rotor time constant, and $\psi_{rq} = 0$. Substituting this expression in the equation (11) yields:

$$\psi_{rd} = \frac{L_m}{1 + sT_r} i_{sd} \tag{12}$$

In the current model the stator flux is represented by:

$$\vec{\psi}_s^i = \frac{L_m}{L_r} \vec{\psi}_r^i + \frac{L_s L_r - L_m^2}{L_r} \vec{i}_s \tag{13}$$

Where $\vec{\psi}_r^i$ is the rotor flux according to the equation (12). Since the voltage model is based on equation (1), the stator flux in the stationary reference frame is given by

$$\vec{\psi}_s = \frac{1}{s}(\vec{v}_s - R_s \vec{i}_s - \vec{U}_{comp}) \tag{14}$$

With the aim to correct the errors associated with the pure integration and the stator resistance measurement, the voltage model is adapted through the PI controller.

$$\vec{U}_{comp} = (K_p + K_i \frac{1}{s})(\vec{\psi}_s - \vec{\psi}_s^i) \tag{15}$$

The K_p and K_i coefficients are calculated with the recommendation proposed in [13]. The rotor flux $\vec{\psi}_r$ in the stationary reference frame is calculated as:

$$\vec{\psi}_r = \frac{L_r}{L_m} \vec{\psi}_s - \frac{L_s L_r - L_m^2}{L_m} \vec{i}_s \tag{16}$$

The estimator scheme shown in the Fig. 2 works with a good performance in the wide range of speeds.

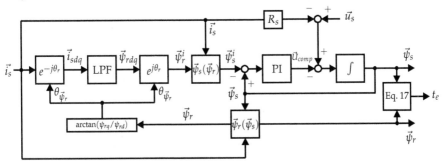

Figure 2. Stator and rotor flux estimator, and electromagnetic torque estimator.

Where LPF means low pass filter. In the other hand, when equations (14) and (16) are replaced in (5) we can estimated the electromagnetic torque t_e as:

$$t_e = \frac{3}{2} P \frac{L_m}{L_r L_s \sigma} \vec{\psi}_r \times \vec{\psi}_s \tag{17}$$

4. Design of fuzzy controllers

4.1. The PI fuzzy controller (PI-F)

The PI fuzzy controller combines two simple fuzzy controllers and a conventional PI controller. Note that fuzzy controllers are responsible for generating the PI parameters dynamically while considering only the torque error variations. The PI-F block diagram is shown in Fig. 3, this controller is composed of two scale factors G_e, $G_{\Delta e}$ at the input. The input for fuzzy controllers are the error (e_N) and error change $(e_{\Delta N})$, and theirs outputs represent the proportional gain K_p and the integral time T_i respectively. These parameters K_p and T_i are adjusted in real time by the fuzzy controllers. The gain K_p is limited to the interval $[K_{p,min}, K_{p,max}]$, which we determined by simulations. For convenience, K_p is normalized in the range between zero and one through the following linear transformation.

$$K'_p = \frac{K_p - K_{p,min}}{K_{p,max} - K_{p,min}} \tag{18}$$

Then, considering that the fuzzy controller output is a normalized value K'_p, we obtain K_p by:

$$K_p = (K_{p,min} - K_{p,max})K'_p + K_{p,min} \tag{19}$$

However, for different reference values the range for the proportional gain values is chosen as $[0, K_{p,max}]$,

$$K_p = K_{p,max}K'_p \tag{20}$$

Due to nonlinearities of the system and in order to avoid overshoots for large reference torque r, it is necessary to reduce the proportional gain. We use a gain coefficient $\rho = 1/(1 + 0002 * r)$ that depends on the reference values. In order to achieves real time adjustment for the K_p values. Therefore, $K_{p,max} = \rho K_{p,max0}$ where the value $K_{p,max0} = 1.24$ was obtained through various simulations. Note that both ρ and $K_{p,max}$ decreases as the reference value increases. Consequently, the gain K_p decreases. The PI-F controller receives as input the torque error e and as output the motor load angle γ^*.

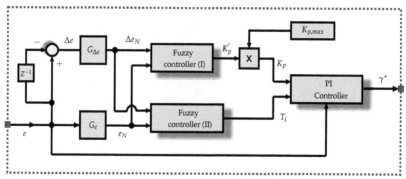

Figure 3. PI Fuzzy Controller block diagram.

4.1.1. Membership Functions (MF)

In the Fig. 3, the first fuzzy controller receives as inputs the errors e_N, Δe_N, each of them has three fuzzy sets that are defined similarly, being only necessary to describe the fuzzy sets of the first input. The first input e_N has three fuzzy sets whose linguistic terms are N-Negative, ZE-Zero and P-Positive. Each fuzzy set has a membership function associated with it. In our particular case of, these fuzzy sets have trapezoidal and triangular shapes as shown in Fig.4. The universe of discourse of these sets is defined over the closed interval $[-1.5, 1.5]$.

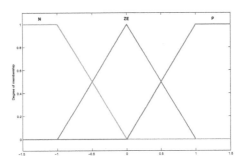

Figure 4. Membership functions for the inputs e_N, Δe_N.

The output has two fuzzy sets whose linguistic values are associated with them are S-small and B-Big, respectively. Both have trapezoidal membership functions as shown in Fig.5. The universe of discourse of the fuzzy sets is defined over the closed interval $[-0.5, 1.5]$. The fuzzy controller uses: triangular norm, Mamdani implication, max-min aggregation method and the center of mass method for defuzzification [11].

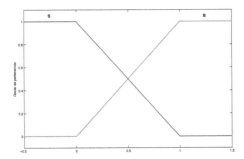

Figure 5. Membership functions for the first fuzzy controller output K_p.

Similarly, the second fuzzy controller has the same fuzzy sets for its two inputs, however, its output is defined by three constant values defined as 1.5, 2 and 3 which linguistic values associated with them are S-Small, M-Medium and B-Big. This controller uses the zero-order Takagi-Sugeno model which simplifies the hardware design and is easy to introduce programmability [19]. The defuzzification method used for this controller is the weighted sum.

4.1.2. Scaling Factors (SF)

The PI-F controller has two scaling factors, G_e and $G_{\Delta e}$ as inputs, while the fuzzy controllers outputs are the gain K_p' and the integral time T_I respectively. From these values we can calculate the parameter $K_I = K_p/T_I$.

The scale factor ensures that both inputs are within the universe of discourse previously defined. The parameters K_p and K_I are the tuned parameters of the PI controller. The inputs are normalized, by:

$$e_N = G_e \cdot e \qquad (21)$$
$$\Delta e_N = G_{\Delta e} \cdot \Delta e \qquad (22)$$

4.1.3. The rule bases

The rules are based on simulation that we conducted of various control schemes. Fig.6 shows an example for one possible response system. Initially, the error is positive around the point a, and the error change is negative, then is imposed a large control signal in order to obtain a small rise time.

To produce a large signal control, the PI controller should have a large gain K_p and a large integral gain K_I (small integral time T_I), therefore,

$$R_x : \text{ if } e_N \text{ is } S \text{ and } \Delta e_N \text{ is } N \text{ then } K_p \text{ is } G$$

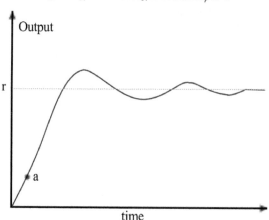

Figure 6. Response system.

The rule base for the first fuzzy controller is in Table 1, also, the rule base for the second fuzzy controller is in Table 2.

Fig. 7 shows the control surface for the first and second fuzzy controllers. This figure clearly shows the non-linear relationship between $(e, \Delta e, K_p)$ and $(e, \Delta e, T_I)$

e_N / Δe_N	N	ZE	P
N	B	B	B
ZE	S	B	S
P	B	B	B

Table 1. Rule base for fuzzy controller (I), output K'_p

e_N / Δe_N	N	ZE	P
N	S	S	S
ZE	B	M	B
P	S	S	S

Table 2. Rule base for fuzzy controller (II), output T_I

4.2. The PI-type fuzzy controller (PIF) and The self-tuning PI-type fuzzy controller (STPIF)

The PI-type fuzzy controller (PIF) is a fuzzy controller inspired by a digital PI controller, which is depicted in Fig. 8. It is composed by two input scale factors "G_e, $G_{\Delta e}$" and one output scale factor "G_{γ^*}". Finally it uses saturation block to limit the output.

This controller has a single input variable, which is the torque error "e" and one output variable which is the motor load angle "γ^*" given by:

$$\gamma^*(k) = \gamma^*(k+1) + \Delta\gamma^*(k) \qquad (23)$$

In (23), k is the sampling time and $\Delta\gamma^*(k)$ represents the incremental change of the controller output. We wish to emphasize here that this accumulation (23) of the controller output takes place out of the fuzzy part of the controller and it does not influence the fuzzy rules.

Fig.9 shows the self-tuning PI-type fuzzy controller (STPIF) block diagram, its main difference with the PIF controller is the gain tuning fuzzy controller (GTF) block.

4.2.1. Membership Functions (MF)

The MF for PIF controller are shown in Fig. 10(a). This MF for input variables "e_N, Δe_N" and output variable "$\Delta\gamma_N^*$" are normalized in the closed interval [-1,1].

The MF's for GTF controller are shown in Fig. 10(a) and in Fig. 10(b) for input and output variables respectively. Input variables "e_N, Δe_N" are defined in the closed interval [-1,1] and the output variable "α" is defined in the closed interval [0,1].

Most of the MF variables have triangular shape [Fig. 10] with 50% overlapping neighbor functions, except the extremes which are trapezoidal. The linguistic variables are referred to as: NL-Negative Large, NM-Negative Medium, NS-Negative Small, ZE-Zero, VS-Very Small, S-Small, SL-Small Large and so on as shown in Table 3 and in Table 4.

4.2.2. Scaling factors

The two inputs SF "$G_{\Delta e}$, G_e" and the output SF "G_{γ^*}" can be adjusted dynamically through updating the scaling factor "α". "α" is computed on-line, using a independent fuzzy rule model

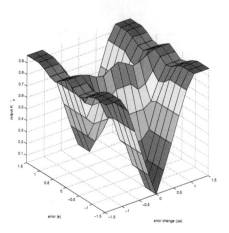

(a) Control surface for fuzzy controller (I), output K'_p.

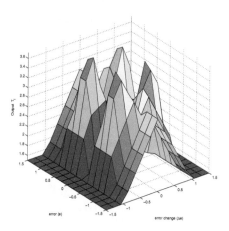

(b) Control surface for fuzzy controller (II), output T_I.

Figure 7. Control surface for: (a) fuzzy controller (I) and (b) fuzzy controller (II)

defined in terms of "$e, \Delta e$". The relationship between the SF and the input/output variables of the STPIF controller are shown bellow:

$$e_N = G_e \cdot e \tag{24}$$

$$\Delta e_N = G_{\Delta e} \cdot \Delta e \tag{25}$$

$$\Delta \gamma^* = (\alpha \cdot G_{\gamma^*}) \cdot \Delta \gamma_N^* \tag{26}$$

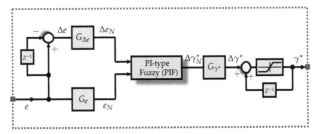

Figure 8. PI-type fuzzy controller.

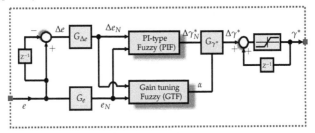

Figure 9. Self-tuning PI-type fuzzy (STPIF) controller.

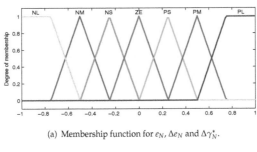

(a) Membership function for e_N, Δe_N and $\Delta \gamma_N^*$.

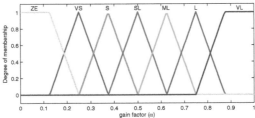

(b) Membership function for α for GTF controller output

Figure 10. Membership functions for: (a) error e_N and change of error Δe_N is the same for PIF and GTF controllers as well as for the output $\Delta \gamma_N^*$ of the PIF controller. (b) the gain updating factor α

4.2.3. The rule bases

The incremental change in the controller output $\Delta\gamma_N^*$ to PIF controller is defined as,

$$R_x : \text{ if } e_N \text{ is } E \text{ and } \Delta e_N \text{ is } \Delta E \text{ then } \Delta\gamma_N^* \text{ is } \Delta\Gamma_N^*$$

Where $E = \Delta E = \Delta\Gamma_N^* = \{NL, NM, NS, ZE, PS, PM, PL\}$. The output α of the GTF controller is determined by the following rules:

$$R_x : \text{ if } e_N \text{ is } E \text{ and } \Delta e_N \text{ is } \Delta E \text{ then } \alpha \text{ is } \chi$$

Where $E = \Delta E = \{NL, NM, NS, ZE, PS, PM, PL\}$ and $\chi = \{ZE, VS, S, SL, ML, L, VL\}$. The rule base for $\Delta\gamma_N^*$ and α are shown in Tab. 3 and Tab. 4 respectively.

Δe_N / e_N	NL	NM	NS	ZE	PS	PM	PL
NL	NL	NL	NL	NM	NS	NS	ZE
NM	NL	NM	NM	NM	NS	ZE	PS
NS	NL	NM	NS	NS	ZE	PS	PM
ZE	NL	NM	NS	ZE	PS	PM	PL
PS	NM	NS	ZE	PS	PS	PM	PL
PM	NS	ZE	PS	PM	PM	PM	PL
PL	ZE	PS	PS	PM	PL	PL	PL

Table 3. Fuzzy rules for the computation of $\Delta\gamma_N^*$

Δe_N / e_N	NL	NM	NS	ZE	PS	PM	PL
NL	VL	VL	VL	L	SL	S	ZE
NM	VL	VL	L	L	ML	S	VS
NS	VL	ML	L	VL	VS	S	VS
ZE	S	SL	ML	ZE	ML	SL	S
PS	VS	S	VS	VL	L	ML	VL
PM	VS	S	ML	L	L	VL	VL
PL	ZE	S	SL	L	VL	VL	VL

Table 4. Fuzzy rules for the computation of α

4.2.4. Gain tuning fuzzy

The purpose of the GTF controller is update continuous the value of α in every sample time. The output α is necessary to control the percentage of the output SF "G_{γ^*}", and therefore for calculating new "$\Delta\gamma^*$",

$$\Delta\gamma^* = (\alpha \cdot G_{\gamma^*}) \cdot \Delta\gamma_N^* \tag{27}$$

The GTF controller rule base is based on knowledge about the three-phase IM control, using a DTC type control according to the scheme proposed in [14], in order to avoid large overshoot and undershoot, e.g., when "e" and "Δe" have different signs, it means that the estimate torque

"t_e" is approaching to the torque reference "t_e^*", then the output SF "G_{γ^*}" must be reduced to a small value by "α", for instance, if "e" is "PM" and "Δe" is "NM" then "α" is "S".

On the other hand, when "e" and "Δe" have the same sign, it means that the torque estimate "t_e" is moving away from the torque reference "t_e^*", the output SF "G_{γ^*}" must be increased to a large value by "α" in order to avoid that the torque depart from the torque reference, e.g., if "e" is "PM" and "Δe" is "PM" then "α" is "VL".

The nonlinear relationship between "$e, \Delta e, \Delta \gamma_N^*$" and "$e, \Delta e, \alpha$" are shown in Fig. 11.

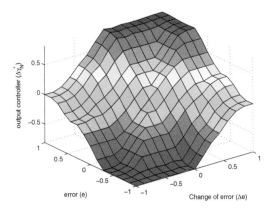

(a) Surface of PIF controller output ($\Delta \gamma_N^*$).

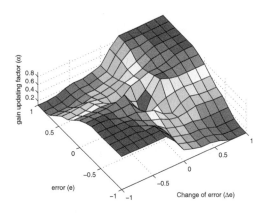

(b) Surface of GTF controller output (α).

Figure 11. Surface of: (a) PIF controller and (b) GTF controller

The inference method used in PIF and GTF controllers is the Mamdani's implication based on max-min aggregation. We use the center of area method for defuzzification.

5. Simulation results

We have conducted our simulation with MATLAB simulation package, which include Simulink block sets and fuzzy logic toolbox. The switching frequency of the pulse width modulation (PWM) inverter was set to be 10kHz, the stator reference flux considered was 0.47 Wb. In order to investigate the effectiveness of the three proposed fuzzy controllers applied in the DTC-SVM scheme we performed several tests.

We used different dynamic operating conditions such as: step change in the motor load (from 0 to 1.0 pu) at 90 percent of rated speed, no-load speed reversion (from 0.5 pu to -0.5 pu) and the application of a specific load torque profile at 90 percent of rated speed. The motor parameters used in the tests are given in Table 5.

Fig. 12, shows the response of the speed and electromagnetic torque when speed reversion for DTC-SVM with PI-F controller is applied. Here, the rotor speed changes its direction at about 1.8 seconds. Fig. 13 shows the stator and rotor current sinusoidal behavior when applying reversion.

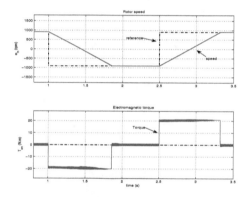

Figure 12. Speed reversion for DTC-SVM with PI-F controller.

Fig. 14 and Fig. 15 show the torque and currents responses respecively, when step change is applied in the motor load for DTC-SVM with the PI-F controller. This speed test was established at 90 percent of rated speed.

In Fig. 16, we demonstrate the speed response when applying a speed reversion for DTC-SVM with PIF controller. In this case the speed of the rotor changes its direction at about 1.4 seconds. Fig. 17 shows the electromagnetic torque behavior when the reversion is applied.

Fig. 18 and Fig. 19 show the response of the electromagnetic torque and phase *a* stator current respectively, when applying a step change in the motor load for DTC-SVM whit PIF controller. In this test the speed of the motor was set to 90 percent of rated speed.

Fig. 20 shows the behaviors of the electromagnetic torque, phase *a* stator current and the motor speed, when applying speed reversion from 0.5 pu to -0.5 pu in the DTC-SVM scheme with STPIF controller. The sinusoidal waveform of the current shows that this control technique also leads to a good current control.

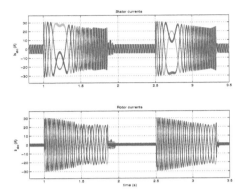

Figure 13. Stator and rotor current for speed reversion for DTC-SVM with PI-F controller.

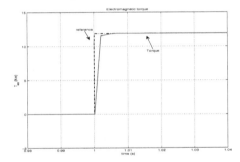

Figure 14. Sudden torque change for DTC-SVM whit PI-F controller

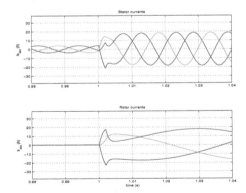

Figure 15. Stator and rotor currents for sudden torque change for DTC-SVM with PI-F controller.

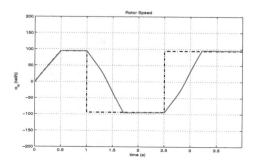

Figure 16. Speed reversion for DTC-SVM with PIF controller.

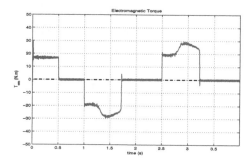

Figure 17. Torque bahavior for speed reversion for DTC-SVM with PIF controller.

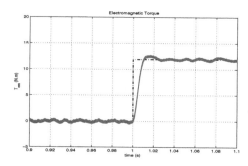

Figure 18. Sudden torque change for DTC-SVM whit PIF controller.

Fig. 21 presents the results when a specific torque profile is imposed to DTC-SVM scheme with STPIF controller. In this case the electromagnetic torque follow the reference.

Fig. 22 illustrates the response of the electromagnetic torque for the DTC-SVM scheme with STPIF controller, when applying step change in the motor load. In this test we used the rise time $t_r = 5.49mS$, the settling time $t_s = 12mS$ and the integral-of-time multiplied by the absolute magnitude of the error index $ITAE = 199.5$.

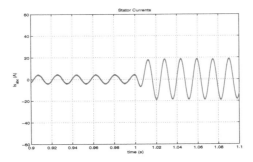

Figure 19. Phase *a* stator current for sudden torque change for DTC-SVM with PIF controller.

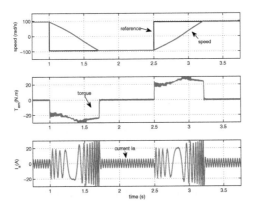

Figure 20. Speed, torque and phase *a* stator current for speed reversion for DTC-SVM with STPIF
controller.

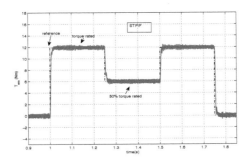

Figure 21. Torque profile for DTC-SVM whit STPIF

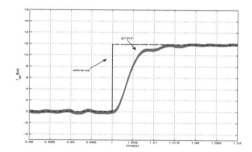

Figure 22. Step change in torque for DTC-SVM scheme with STPIF

Rated voltage (V)	220/60Hz
Rated Power (W)	2238
Rated Torque (Nm)	11.9
Rated Speed (rad/s)	179
$R_s, R_r\,(\Omega)$	0.435, 0.816
L_{ls}, L_{lr} (H)	0.002, 0.002
L_m (H)	0.0693
$J(K_g m^2)$	0.089
P	2

Table 5. Induction Motor Parameters [12]

6. Conclusion

In this chapter we have presented the DTC-SVM scheme that controls a three-phase IM using three different kinds of fuzzy controllers. These fuzzy controllers were used in order to determinate dynamically and on-line the load angle between stator and rotor flux vectors. Therefore, we determine the electromagnetic torque necessary to supply the motor load. We have conducted simulations with different operating conditions. Our simulation results show that the all proposed fuzzy controllers work appropriately and according to the schemes reported in the literature. However, the STPIF controller achieves a fast torque response and low torque ripple in a wide range of operating conditions such as: sudden change in the command speed and step change of the load.

Acknowledgements

The authors are grateful to FAPESP and CAPES for partially financial support.

Author details

José Luis Azcue and Ernesto Ruppert
School of Electrical and Computer Engineering of University of Campinas, UNICAMP, Department of Energy Control and Systems, Campinas-SP, Brazil

Alfeu J. Sguarezi Filho
CECS/UFABC, Santo André - SP, Brazil

7. References

[1] Abu-Rub, H., Guzinski, J., Krzeminski, Z. & Toliyat, H. [2004]. Advanced control of induction motor based on load angle estimation, *Industrial Electronics, IEEE Transactions on* 51(1): 5 – 14.

[2] Azcue P., J. L. [2010]. *Three-phase induction motor direct torque control using self-tuning pi type fuzzy controller*, Master's thesis, University of Campinas (UNICAMP).

[3] Azcue P., J. & Ruppert, E. [2010]. Three-phase induction motor dtc-svm scheme with self-tuning pi-type fuzzy controller, *Fuzzy Systems and Knowledge Discovery (FSKD), 2010 Seventh International Conference on*, Vol. 2, pp. 757 –762.

[4] Bertoluzzo, M., Buja, G. & Menis, R. [2007]. A direct torque control scheme for induction motor drives using the current model flux estimation, *Diagnostics for Electric Machines, Power Electronics and Drives, 2007. SDEMPED 2007. IEEE International Symposium on* pp. 185 –190.

[5] Cao, S.-G., Rees, N. & Feng, G. [1999]. Analysis and design of fuzzy control systems using dynamic fuzzy-state space models, *Fuzzy Systems, IEEE Transactions on* 7(2): 192 –200.

[6] Chen, C.-L. & Chang, M.-H. [1998]. Optimal design of fuzzy sliding-mode control: A comparative study, *Fuzzy Sets and Systems* 93(1): 37 – 48.
URL: *http://www.sciencedirect.com/science/article/pii/S0165011496002217*

[7] Chen, S., Kai, T., Tsuji, M., Hamasaki, S. & Yamada, E. [2005]. Improvement of dynamic characteristic for sensorless vector-controlled induction motor system with adaptive pi mechanism, *Electrical Machines and Systems, 2005. ICEMS 2005. Proceedings of the Eighth International Conference on*, Vol. 3, pp. 1877 –1881 Vol. 3.

[8] Depenbrock, M. [1988]. Direct self-control (dsc) of inverter-fed induction machine, *Power Electronics, IEEE Transactions on* 3(4): 420–429.

[9] Habetler, T., Profumo, F., Pastorelli, M. & Tolbert, L. [1992]. Direct torque control of induction machines using space vector modulation, *Industry Applications, IEEE Transactions on* 28(5): 1045 –1053.

[10] Koutsogiannis, Z., Adamidis, G. & Fyntanakis, A. [2007]. Direct torque control using space vector modulation and dynamic performance of the drive, via a fuzzy logic controller for speed regulation, *Power Electronics and Applications, 2007 European Conference on* pp. 1 –10.

[11] Kovacic, Z. & Bogdan, S. [2006]. *Fuzzy Controller Design: Theory and Applications*, CRC Press.

[12] Krause, P. C., Wasynczuk, O. & Sudhoff, S. D. [2002]. *Analysis of Electric Machinery and Drive Systems*, IEEE Press.

[13] Lascu, C., Boldea, I. & Blaabjerg, F. [2000]. A modified direct torque control for induction motor sensorless drive, *Industry Applications, IEEE Transactions on* 36(1): 122–130.

[14] Rodriguez, J., Pontt, J., Silva, C., Kouro, S. & Miranda, H. [2004]. A novel direct torque control scheme for induction machines with space vector modulation, *Power Electronics Specialists Conference, 2004. PESC 04. 2004 IEEE 35th Annual*, Vol. 2, pp. 1392 – 1397 Vol.2.

[15] Takahashi, I. & Noguchi, T. [1986]. A new quick-response and high-efficiency control strategy of an induction motor, *Industry Applications, IEEE Transactions on* IA-22(5): 820 –827.

[16] Trzynadlowski, A. M. [1993]. *The Field Orientation Principle in Control of Induction Motors*, springer.

[17] Vas, P. [1998]. *Sensorless Vector and Direct Torque Control*, Oxford University Press. ISBN 0198564651.

[18] Viola, J., Restrepo, J., Guzman, V. & Gimenez, M. [2006]. Direct torque control of induction motors using a fuzzy inference system for reduced ripple torque and current limitation, *Power Electronics and Motion Control Conference, 2006. EPE-PEMC 2006. 12th International*, pp. 1161 –1166.

[19] Yamakawa, T. [1993]. A fuzzy inference engine in nonlinear analog mode and its application to a fuzzy logic control, *Neural Networks, IEEE Transactions on* 4(3): 496 –522.

Permissions

The contributors of this book come from diverse backgrounds, making this book a truly international effort. This book will bring forth new frontiers with its revolutionizing research information and detailed analysis of the nascent developments around the world.

We would like to thank Sohail Iqbal, Nora Boumella and Juan Carlos Figueroa-García, for lending their expertise to make the book truly unique. They have played a crucial role in the development of this book. Without their invaluable contribution this book wouldn't have been possible. They have made vital efforts to compile up to date information on the varied aspects of this subject to make this book a valuable addition to the collection of many professionals and students.

This book was conceptualized with the vision of imparting up-to-date information and advanced data in this field. To ensure the same, a matchless editorial board was set up. Every individual on the board went through rigorous rounds of assessment to prove their worth. After which they invested a large part of their time researching and compiling the most relevant data for our readers. Conferences and sessions were held from time to time between the editorial board and the contributing authors to present the data in the most comprehensible form. The editorial team has worked tirelessly to provide valuable and valid information to help people across the globe.

Every chapter published in this book has been scrutinized by our experts. Their significance has been extensively debated. The topics covered herein carry significant findings which will fuel the growth of the discipline. They may even be implemented as practical applications or may be referred to as a beginning point for another development. Chapters in this book were first published by InTech; hereby published with permission under the Creative Commons Attribution License or equivalent.

The editorial board has been involved in producing this book since its inception. They have spent rigorous hours researching and exploring the diverse topics which have resulted in the successful publishing of this book. They have passed on their knowledge of decades through this book. To expedite this challenging task, the publisher supported the team at every step. A small team of assistant editors was also appointed to further simplify the editing procedure and attain best results for the readers.

Our editorial team has been hand-picked from every corner of the world. Their multi-ethnicity adds dynamic inputs to the discussions which result in innovative outcomes. These outcomes are then further discussed with the researchers and contributors who give their valuable feedback and opinion regarding the same. The feedback is then collaborated with the researches and they are edited in a comprehensive manner to aid the understanding of the subject.

Apart from the editorial board, the designing team has also invested a significant amount of their time in understanding the subject and creating the most relevant covers. They scrutinized every image to scout for the most suitable representation of the subject and create an appropriate cover for the book.

The publishing team has been involved in this book since its early stages. They were actively engaged in every process, be it collecting the data, connecting with the contributors or procuring relevant information. The team has been an ardent support to the editorial, designing and production team. Their endless efforts to recruit the best for this project, has resulted in the accomplishment of this book. They are a veteran in the field of academics and their pool of knowledge is as vast as their experience in printing. Their expertise and guidance has proved useful at every step. Their uncompromising quality standards have made this book an exceptional effort. Their encouragement from time to time has been an inspiration for everyone.

The publisher and the editorial board hope that this book will prove to be a valuable piece of knowledge for researchers, students, practitioners and scholars across the globe.

List of Contributors

Nora Boumella
University of Batna, Batna - Algeria

Juan Carlos Figueroa
Universidad Distrital Francisco Jose de Caldas, Bogota - Colombia

Sohail Iqbal
NUST-SEECS, Islamabad – Pakistan

Carlos André Guerra Fonseca
Informatics Department, Rio Grande do Norte State University, Natal, Brazil
Computing Engineering and Automation Department, Rio Grande do Norte Federal
University, Natal, Brazil

Fábio Meneghetti Ugulino de Araújo
Computing Engineering and Automation Department, Rio Grande do Norte Federal
University, Natal, Brazil

Marconi Câmara Rodrigues
Science and Technology School, Rio Grande do Norte Federal University, Natal, Brazil

Morteza Seidi, Marzieh Hajiaghamemar and Bruce Segee
University of Maine, USA

Muhammad M.A.S. Mahmoud
Received the B.S. degree in Electrical Engineering from Cairo University and the M.Sc.
degree from Kuwait University, PH.D Transilvania University of Brasov, Romania
He occupies a position of Senior Engineer at Al Hosn Gas Co. UAE

Wudhichai Assawinchaichote
Department of Electronic and Telecommunication Engineering at King Mongkut's University of Technology Thonburi, Bangkok, Thailand

Yassine Manai and Mohamed Benrejeb
National Engineering School of Tunis, LR-Automatique, Tunis, Tunisia

Georgios A. Tsengenes and Georgios A. Adamidis
Department of Electrical and Computer Engineering, Democritus University of Thrace,
Greece

Mavungu Masiala, Mohsen Ghribi and Azeddine Kaddouri
Greater Research Group, University of Moncton, New-Brunswick, Canada

B. S. K. K. Ibrahim
Dept of Mechatronics and Robotics, Faculty of Electrical & Electronic Engineering, University Tun Hussein Onn Malaysia, Batu Pahat, Johor, Malaysia

M. O. Tokhi, M. S. Huq and S. C. Gharooni
Department of Automatic Control and System Engineering, University of Sheffield, United Kingdom

José Luis Azcue and Ernesto Ruppert
School of Electrical and Computer Engineering of University of Campinas, UNICAMP, Department of Energy Control and Systems, Campinas-SP, Brazil

Alfeu J. Sguarezi Filho
CECS/UFABC, Santo André - SP, Brazil

Printed in the USA
CPSIA information can be obtained
at www.ICGtesting.com
JSHW011813301024
72690JS00002B/72